T0224070

Communications
in Computer and Information Science   1131

Commenced Publication in 2007
Founding and Former Series Editors:
Phoebe Chen, Alfredo Cuzzocrea, Xiaoyong Du, Orhun Kara, Ting Liu,
Krishna M. Sivalingam, Dominik Ślęzak, Takashi Washio, Xiaokang Yang,
and Junsong Yuan

Editorial Board Members

Jerry Chun-Wei Lin · I-Hsien Ting ·
Tiffany Tang · Kai Wang (Eds.)

Multidisciplinary
Social Networks Research

6th International Conference, MISNC 2019
Wenzhou, China, August 26–28, 2019
Revised Selected Papers

 Springer

Editors
Jerry Chun-Wei Lin
Western Norway University
of Applied Sciences
Bergen, Norway

I-Hsien Ting
Department of Information Management
National University of Kaohsiung
Kaohsiung, Taiwan

Tiffany Tang
Wenzhou University
Wenzhou, China

Kai Wang
Department of Information Management
National University of Kaohsiung
Kaohsiung, Taiwan

ISSN 1865-0929   ISSN 1865-0937 (electronic)
Communications in Computer and Information Science
ISBN 978-981-15-1757-0   ISBN 978-981-15-1758-7 (eBook)
https://doi.org/10.1007/978-981-15-1758-7

This Springer imprint is published by the registered company Springer Nature Singapore Pte Ltd.
The registered company address is: 152 Beach Road, #21-01/04 Gateway East, Singapore 189721, Singapore

Preface

The 6th Multidisciplinary International Social Networks Conference (MISNC 2019) was held during August 26–28, 2019, in Wenzhou, China.

We welcomed all participants of MISNC 2019 at Wenzhou-Kean University, China. The conference, in its 6th year, grows globally. In 2014, the MISNC conference was hosted in Kaohsiung, Taiwan. The following conferences were held in 2015 (Matsuyama, Japan), 2016 (Kean University, New Jersey, USA), 2017 (Bangkok, Thailand), and 2018 (St. Etienne, France). This year, we had the pleasure to socialize in authentic Chinese culture and network with international researchers, professionals, experts, and students from all over the world.

We are proud to report that there were participants from seven countries/areas, including Japan, Taiwan, China, USA, India, Hong Kong, and Norway. We also invited two keynote speakers from Taiwan and the USA. This year, we received 37 submissions, and 15 high-quality papers were accepted to be presented at the conference.

The conference would not have been possible without our partners and sponsors: Wenzhou-Kean University, the National University of Kaohsiung and its Social Network Innovation Center (SNIC), the Ministry of Science and Technology of Taiwan, the Taiwanese Association for Social Networks (TASN), the Western Norway University of Applied Sciences, and IEEE SMC Tainan Section.

Last and most importantly, we thank all of you, the authors and attendees for having participated in MISNC 2019, sharing the knowledge and experience, and contributing to the advancement of science and technology for the improvement of the quality of our lives. We hope that each of you had a most pleasant experience at MISNC 2019. We also look forward to seeing you in MISNC 2020 in Okinawa.

August 2019

Jerry Chun-Wei Lin
I-Hsien Ting
Tiffany Tang
Kai Wang

Organization

Honorary Chairs

Leon S. L. Wang   National University of Kaohsiung, Taiwan
Tzung-Pei Hong   National University of Kaohsiung, Taiwan

General Chair

I-Hsien Ting   National University of Kaohsiung, Taiwan

Program Chairs

Jerry Chun-Wei Lin   Western Norway University of Applied Science,
     Norway
Tiffany Y. Tang   Wenzhou-Kean University, China
Kai Wang   National University of Kaohsiung, Taiwan

Event Co-chair

Fa-Hsiang (Jacob) Chang   Wenzhou-Kean University, China

Publication Chair

Hsin-Chang Yang   National University of Kaohsiung, Taiwan

Publicity Chairs

Been-Chian Chien   National Tainan University, Taiwan
Georgios Lappas   Technological Educational Institute of Western
     Macedonia, Greece
Qian Li   University of Technology Sydney, Australia
Andi Mursidi   STKIP Singkawang, Indonesia
Didi Sundiman   Universitas Universal, Indonesia

Special Session Chair

Tzu-Hsien Yang   National University of Kaohsiung, Taiwan

Program Committee Members

Prantik Bhattacharyya   Reddit, USA
Dan Braha   NECSI, UK

Piotr Brodka	Wroclaw University of Science and Technology, Poland
Bao-Rong Chang	National University of Kaohsiung, Taiwan
Chien-Chung Chan	University of Akron, USA
Fa-Hsiang (Jacob) Chang	Wenzhou-Kean University, China
Wei-Lun Chang	Tamkang University, Taiwan
Richard Chbeir	LIUPPA Laboratory, France
Rongjuan Chen	Wenzhou-Kean University, China
Shin-Horng Chen	Southern Taiwan University of Science and Technology, Taiwan
Yi-Chung Chen	Feng Chia University, Taiwan
Yunwei Chen	Chengdu Library of Chinese Acad Sci, China
Been-Chian Chien	National Tainan University, Taiwan
Candy Lim Chiu	Wenzhou-Kean University, China
Vicente Garcia Diaz	University of Oviedo, Spain
Paolo Garza	Politecnico di Torino, Italy
Juan Carlos Figueroa-Garcia	Universidad Distrital Francisco Jose de Caldas, Colombia
William Grosky	University of Michigan, USA
Han-Chiang Ho	Wenzhou-Kean University, China
Pei-Lin Hsu	Oriental Institute of Technology, Taiwan
Han-Fen Hu	University of Nevada, Las Vegas, USA
Pin-Rui Hwang	National United University, Taiwan
Jason Jung	Cung-Ang University, South Korea
Chutisant Kerdvibulvech	National Institute of Development Administration (NIDA), Thailand
Ying-Feng Kuo	National University of Kaohsiung, Taiwan
Cheng-Yu Lai	Chung Yuan Christian University, Taiwan
Y. H. Lai	Oriental Institute of Technology, Taiwan
Georgios Lappas	Technological Educational Institute of Western Macedonia, Greece
Gang Li	Deakin University, Australia
Paoling Liao	National Kaohsiung University of Science and Technology, Taiwan
Jerry Chun-Wei Lin	Western Norway University of Applied Sciences, Norway
João Cordeiro	Catholic University of Portugal, Portugal
Wen-Yang Lin	National University of Kaohsiung, Taiwan
Andi Mursidi	STKIP Singkawang, Indonesia
Ben Daniel Motidyang	University of Otago, New Zealand
Sanetake Nagayoshi	Shizuoka University, Japan
Takashi Okamoto	Ehime University, Japan
Kok-Leong Ong	La Trobe University, Australia
Fatih Ozgul	University of Hertfordshire, UK
Suppanunta Romprasert	Srinakharinwirot University, Thailand
Laura Pullum	Oak Ridge National Laboratory, USA

Deffi Ayu Puspito Sari	Universitas Bakrie, Indonesia
Hidenobu Sai	Ehime University, Japan
Muhammad Ikhsan Setiawan	Narotama University, Indonesia
Rong-An Shang	Soochow University, Taiwan
Norihito Seki	Hokkai-Gakuen University, Japan
Jun Shen	University of Wollongong, Australia
Didi Sundiman	Universitas Universal, Indonesia
Andrea Tagarelli	University of Calabria, Italy
Xiaohui Tao	University of Southern Queensland, Australia
I-Hsien Ting	National University of Kaohsiung, Taiwan
Masatsugu Tsuji	Kobe International University, Japan
Yasushi Ueki	Institute of Developing Economies, Japan
Shiro Uesugi	Matsuyama University, Taiwan
Li Weigang	University of Brasília, Brazil
Kai Wang	National University of Kaohsiung, Taiwan
Kai-Yu Wang	Brock University, Canada
Leon Wang	National University of Kaohsiung, Taiwan
Uffe Wiil	University of Southern Denmark, Denmark
Pinata Winoto	Wenzhou-Kean University, China
Chienhsing Wu	National University of Kaohsiung, Taiwan
Chuan-Chun Wu	I-Shou University, Taiwan
Hsin-Chang Yang	National University of Kaohsiung, Taiwan
Shu-Chen Yang	National University of Kaohsiung, Taiwan
Xingquan Zhu	Florida Atlantic University, USA
Yi Zuo	Nagoya University, Japan

Contents

Supervised Deep Learning for Hierarchical Image Data Retrieval

Been-Chian Chien[✉], Yueh-Chia Hsu, and Ya-Yu Huang

Department of Computer Science and Information Engineering,
National University of Tainan, Tainan, Taiwan, R.O.C.
bcchien@mail.nutn.edu.tw, maryzxc12345@gmail.com,
yayuhuang36@gmail.com

Abstract. The techniques of feature extraction and representation on image data have been significantly progressed in recent years due to the development of deep learning. With a large number of representative image features being extracted from ImageNet by convolutional neural networks, many object recognizing applications were successfully accomplished effectively. In this paper, two supervised image retrieval models for retrieving images with similar hierarchical concept are investigated and compared. First, image features are extracted by pre-trained VGG convolutional networks. Then, the supervised retrieval models are learned from a set of images with hierarchical concept labels. The experimental results show that the hash-based model generally is superior to classifier-based model both in F1 measure and MAP no matter what in coarse level or fine level of concept hierarchy.

Keywords: Convolutional neural network · Image retrieval · Image classification · Supervised deep learning · Image concept hierarchy

1 Introduction

With the extremely increasing of image data on web of late years, it is a challenge for researchers to retrieve the relevant images according to users' semantic request from a large-scale image data set effectively and efficiently. Content-based image retrieval and keyword-based image retrieval are the two typical methodologies nowadays used to help retrieving relevant images for users. A content-based image retrieval system must give an example image to the system for searching the images with relevant visual content. If users cannot provide the appropriate target example, the system may not be able to return satisfying retrieval results. Instead of giving image examples, a keyword-based image retrieval system needs provide semantic keywords only. However, the drawback of this approach is that images in the database have to be annotated by texts in advance. The images cannot be effectively retrieved if the annotated keyword is missing or the describing text is different from the relevant images.

For a content-based image retrieval system, feature extraction, similarity measuring and feature indexing are the main research issues, but accurately annotating concept for images is the key of supporting effective keyword-based image retrieval. Supervised learning for image classification can help annotating images from a finite set of concept

© Springer Nature Singapore Pte Ltd. 2019
J. C.-W. Lin et al. (Eds.): MISNC 2019, CCIS 1131, pp. 1–13, 2019.
https://doi.org/10.1007/978-981-15-1758-7_1

automatically. As a matter of fact, feature extraction and similarity measuring are also the essential problems in image classification. In the past, color, shape and texture are conventional features used to represent an image. However, these features did not gain much improvement in image retrieval and image classification. In recent years, convolutional neural network (CNN) of deep learning was proposed for extracting image feature representation from a given set of images [4]. The recognition rate in the task of image classification and annotation thus is greatly improved by the extracted image features using convolutional networks. The effectiveness of image retrieval is also improved since the image features extracted by convolutional networks can capture semantic of images more precisely.

Although convolutional networks are capable of extracting better image representation, image features generally are represented as high-dimensional real values. The computation of similarity measure in a large-scale dataset with millions of images still needs to cost a lot of time and storage. To resolve the inefficiency of computing real-valued features, learning-to-hash approaches were proposed to find similarity-preserving binary codes to accomplish approximate nearest neighbor search. Further, most of the work on learning binary hash codes aim at the generation of effective binary codes for flat image concept. In the paper, we investigate and compare two supervised image retrieval models based on deep learning feature extraction for retrieving images with similar hierarchical semantic concept. First, image features are extracted by VGG convolutional networks pre-trained from ImageNet. After extracting image features, two supervised image retrieval models, the classifier-based retrieval and the hash-based retrieval, are learned from a set of images labeled in a semantic concept hierarchy. The two models are tested and compared by PASCAL VOC 2012 dataset. The experimental results show that the hash-based model generally is superior to classifier-based model both in F1 measure and MAP no matter what in coarse level or fine level of concept hierarchy.

The remainder of this paper is organized as follows. Section 2 reviews the related research of image representation and learning of binary hash codes. Section 3 describes the structure of image concept hierarchy and the two learning models for semantic image retrieval. The experimental results are shown and discussed in Sect. 4. Finally, a concluding remark and future work are given in Sect. 5.

2   Related Works

The deep learning of convolutional neural networks was first well-known as AlexNet by researchers in the ISLVRC 2012 image recognition competition of ImageNet [4]. From then on, many improved convolutional networks were proposed to seek better architecture for image recognition, such as VGGNet [7], GoogLeNet [8], etc. Automatic image features learning and extraction are the main superiority of convolutional networks. The advantage of feature extraction is also applied and discussed for constructing effective image retrieval [10, 12].

On large-scale image datasets, the similarity computation and feature indexing are important for building an efficient image retrieval system. Unsupervised hashing methods like LSH (locality sensitive hashing) [3] try to encode image features to binary codes preserving the similarity of original features. However, retrieval performance is still subject to the semantic gap between low-level image features and high-level labeling. Of late years, supervised learning hash code methods are developed to fetch up the linking information between image features and semantic labels. Xia et al. [9] proposed the method CNNH using a two-stage strategy to map images to hash codes. A number of CNN-based hashing methods [5, 6, 13] were proposed to learn image features together with binary hash codes. DHN [15] further applies a cross-entropy loss and a quantization loss to preserve the image pair-wise similarity and control the quantization error. HashNet [1] presents a new architecture for deep learning to hash to handle the problem of binary code convergence. With the continuation method, deep feature learning and binary hash coding can be accomplished end-to-end in an integrated network.

Some researches of classifying and retrieval on hierarchical image data are also proposed in recent years. A CNN-RNN architecture is proposed to demonstrate its performance in hierarchical image classification by Guo et al. [2]. The supervised deep hashing method is also extend to learn binary hash codes for hierarchical image retrieval in [11, 14].

3 The Approaches and Models

3.1 Hierarchical Image Structures

Generally, each image in the image datasets will be assigned a label or multi-labels for the problems of image classification and image retrieval. For a hierarchical image dataset, image labels are usually organized as a meaningful semantic concept hierarchy. The hierarchical image data is further described as the following definitions.

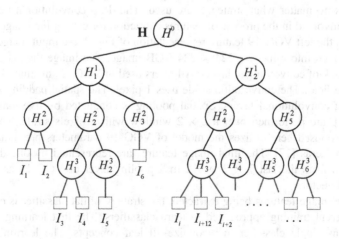

Fig. 1. An example of concept hierarchy for image data.

A concept hierarchy $\mathbf{H}$ is a tree-like structure with a root H^0 as the top level representing the most general concept in the hierarchy. H_j^i stands for the jth concept at concept level i, where $i, j \geq 1$. The concept H_k^{i+1} is called a child concept of H_j^i and H_j^i is called the parent concept of H_k^{i+1} if H_k^{i+1} is a direct sub-concept of H_j^i. A leaf concept is a concept H_j^i without any child concept. Let $\mathbf{L}$ be denoted as the set of all leaf concepts in $\mathbf{H}$. For an image data set $\mathbf{I}$, assume that $\mathbf{y}_i$ is the set of leaf concept labels for images $I_i \in \mathbf{I}$, $\mathbf{y}_i \subset \mathbf{L}$; if an image I_i is labeled as the concept $H_j^i \in \mathbf{y}_i$, I_i is also in the all ancestor concepts of H_j^i.

As the example concept hierarchy $\mathbf{H}$ in Fig. 1, H^0 is the root level. There are two concepts H_1^1 and H_2^1 in concept level one. Level two has five concepts from H_1^2 and H_5^2. The total number of concept levels is three. The set of leaf concepts $\mathbf{L} = \{H_1^2, H_3^2, H_1^3, H_2^3, H_3^3, H_4^3, H_5^3, H_6^2\}$. The set of concept labels for image I_5 is $\{H_2^2\}$, and it means that the concepts H_2^2 and H_1^1 also contain the image I_5.

3.2   The Classifier-Based Models

Traditional supervised strategies usually collect representative image datasets to learn classifiers for retrieving images with similar concept. A classifier may be enough while the target concepts are flat and evident. However, most of images imply more than one concept inside and are formed in co-existing or hierarchical formats. A shallow classifier generally cannot archive effective recognition for complicated concept structures. Instead of single classifier, the strategy of generating a set of classifiers is considered to tackle such a task. For comparing the effectiveness of image retrieval, two classification strategies based on random forest classifiers are employed to identify the possible concepts for hierarchical image datasets. The first strategy applies flat classifiers, and the other exploits the strategy of local classifier per parent node.

The first step of building an image concept retrieving system is to extract features from images no matter what strategies are used. The deep convolutional network of VGG [7] is involved in the process of features extraction for getting the image features. As shown in the left VGG-19 feature representation of Fig. 2, the input image must be first transformed into a fix-size 224×224 RGB image. The image then is processed through a stack of convolutional layers. All filters used in VGG are in small 3×3 size of respective field. The convolution stride uses 1 pixel. The spatial padding is 1 pixel for 3×3 of convolutional layers. Spatial pooling is conducted by five max-pooling layers, which are performed over a 2×2 window with 2 pixels stride. The feature extraction proposed here utilizes the model of VGG-19 parameters pre-trained using ImageNet in Keras [17]. The final image features are represented by one dimension vector flatten from the last $7 \times 7 \times 512$ max-pooling layer of VGG-19 as a 25,088 dimensional vector.

After the image features being extracted, the strategy of flat classifier is conducted by two different training approached to learn classifiers. The first learning approach directly trains single classifier to recognize all leaf concepts. The learning process denoted as RF-1 is shown as follows.

Learner RF-1
{
    for all $(I_i \in \mathbf{I})$
       $x_i \leftarrow \text{VGG19}(I_i)$;
    $\mathbf{T} = \{(x_i, y_i) \mid y_i \subset \mathbf{L}\}$;
    RF-1 $\leftarrow$ {RandomForest_train($\mathbf{T}$)} ;
}

The second learning approach respectively builds an individual classifier for each leaf concept. The learning process named RF-n is shown as follows.

Learner RF-n
{
    for $(I_i \in \mathbf{I})$
       $x_i \leftarrow \text{VGG19}(I_i)$;
    RF-$n = \varnothing$;
    for each $H^k_j \in \mathbf{L}$
    {
      for $(I_i \in \mathbf{I})$
        if $(H^k_j \in y_i)$
          $\mathbf{T}^+ = \mathbf{T}^+ \cup \{(x_i, H^k_j)\}$;
        else
          $\mathbf{T}^- = \mathbf{T}^- \cup \{(x_i, \sim H^k_j)\}$;
      RF-$n \leftarrow$ RF-$n \cup$ {RandomForest_train($\mathbf{T}^+ \cup \mathbf{T}^-$)} ;
    }
}

Since the flat classifier strategy identifies the leaf concepts, the higher concepts in concept hierarchy will be determined at the same time. On the contrary, for the strategy of local classifier per parent node trains local classifiers for non-leaf concepts with more than two sub-concepts. This model combines multiple classifiers is called mRF. The training data sets and learning process are performed as follows.

Learner mRF
{
    for $(I_i \in \mathbf{I})$
       $x_i \leftarrow \text{VGG19}(I_i)$;
    mRF $= \varnothing$;
    for each $H^k_j \notin \mathbf{L}$
    {
      for each $H \in \mathbf{L}$
        $\mathbf{T}_k = \{(x_i, child(H^k_j)) \mid H \in y_i \text{ and } H \in sub\_concept(H^k_j)\}$;
      mRF $\leftarrow$ mRF $\cup$ {RandomForest_train($\mathbf{T}_k$)} ;
    }
}

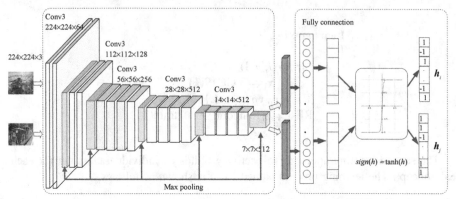

VGG-19 feature representation

Fig. 2. The deep convolutional networks for learning hash codes.

3.3 The Hash-Based Model

The hash-based code learning approach is the other scheme for efficient similarity image retrieval. The HashNet [1] are applied to produce binary hash codes for representing Hamming distance between two image features. The supervised deep hashing model is shown as Fig. 2. The first part of feature extraction also adopts VGG-19 convolutional networks to extract image features. Let $I_i, I_j \in \mathbf{I}$, if I_i and I_j are labeled as the same leaf concept, the similarity of I_i and I_j will be marked as $z_{ij} = 1$; otherwise, $z_{ij} = 0$. Assume that the $\mathbf{x}_i$ and $\mathbf{x}_j$ are the image features of images I_i and I_j extracted through the convolutional networks layer of VGGNets. The goal of learning to hash is to learn a hash function h for the set of image pairs $\{(\mathbf{x}_i, \mathbf{x}_j, z_{ij}) \mid \text{for } I_i, I_j \in \mathbf{I}\}$, $h : (\mathbf{x}_i, \mathbf{x}_j) \rightarrow (\mathbf{h}_i, \mathbf{h}_j)$, where $\mathbf{h}_i, \mathbf{h}_j \in \{-1, 1\}^K$ are respectively K-bits binary codes. In order to train binary codes from continuous activation function, the output code uses -1 to replace the symbol 0. Let $\langle \mathbf{h}_i \cdot \mathbf{h}_j \rangle$ denote the inner product of $\mathbf{h}_i, \mathbf{h}_j$, thus Hamming distance between the codes $\mathbf{h}_i, \mathbf{h}_j, d_H(\mathbf{h}_i, \mathbf{h}_j)$, can be defined as

$$d_H(\mathbf{h}_i, \mathbf{h}_j) = \frac{1}{2}(K - \langle \mathbf{h}_i \cdot \mathbf{h}_j \rangle). \tag{1}$$

To preserve the similarity of images for the corresponding hash codes $\{h_i| \text{ for } I_i \in \mathbf{I}\}$, a conditional probability $P(z_{ij}|\mathbf{h}_i, \mathbf{h}_j)$ was introduced to be the base of objective likelihood estimation. The more similar the binary code pair $\mathbf{h}_i, \mathbf{h}_j$ is, the smaller the Hamming distance $d_H(\mathbf{h}_i, \mathbf{h}_j)$ is, at the same time, and the larger the inner product $\langle \mathbf{h}_i \cdot \mathbf{h}_j \rangle$ is. Thus, the probability of $P(1|\mathbf{h}_i, \mathbf{h}_j)$ should have a larger value while the inner product $\langle \mathbf{h}_i \cdot \mathbf{h}_j \rangle$ is large; on the contrary, $P(0|\mathbf{h}_i, \mathbf{h}_j)$ will be large if the inner product $\langle \mathbf{h}_i \cdot \mathbf{h}_j \rangle$ is small. Let

$$P(z_{ij}|\boldsymbol{h}_i, \boldsymbol{h}_j) = \begin{cases} \sigma(\langle \boldsymbol{h}_i \cdot \boldsymbol{h}_j \rangle), & z_{ij} = 1; \\ 1 - \sigma(\langle \boldsymbol{h}_i \cdot \boldsymbol{h}_j \rangle), & z_{ij} = 0; \end{cases} \tag{2}$$

where $\sigma(x) = 1/(1 + e^{-\alpha x})$ is the sigmoid function with an adaptive hyper-parameter α, and generally we set $\alpha \leq 1$.

For adapting to the problem of data imbalance between similar and dissimilar in multiple classes, the weight w_{ij} is defined as follows,

$$w_{ij} = \frac{|\boldsymbol{y}_i \cap \boldsymbol{y}_j|}{|\boldsymbol{y}_i \cup \boldsymbol{y}_j|} \cdot \begin{cases} |\boldsymbol{Z}|/|\boldsymbol{Z}_1|, & z_{ij} = 1; \\ |\boldsymbol{Z}|/|\boldsymbol{Z}_0|, & z_{ij} = 0; \end{cases} \tag{3}$$

where $\boldsymbol{Z} = \{z_{ij}|$ for all $I_i, I_j \in \boldsymbol{I}\}, \boldsymbol{Z}_1 = \{z_{ij}|z_{ij} = 1, z_{ij} \in \boldsymbol{Z}\}, \boldsymbol{Z}_0 = \{z_{ij}|z_{ij} = 0, z_{ij} \in \boldsymbol{Z}\}$.

By combining the both consideration of Eqs. (2) and (3), the weighted maximum likelihood estimation is established. For the training dataset $\{(x_i, x_j, z_{ij}) \mid$ for $I_i, I_j \in \boldsymbol{I}\}\}$, to archive the learning optimization objective of generating $h : (x_i, x_j) \rightarrow (\boldsymbol{h}_i, \boldsymbol{h}_j)$, the following loss function is applied.

$$\min_{\Theta} \sum_{z_{ij} \in \boldsymbol{Z}} w_{ij}(\log(1 + \exp(\alpha\langle \boldsymbol{h}_i \cdot \boldsymbol{h}_j \rangle))) - z_{ij}\alpha\langle \boldsymbol{h}_i \cdot \boldsymbol{h}_j \rangle) \tag{4}$$

where Θ is the set of all possible parameters in the deep learning networks as Fig. 2. In fact, the direct outputs in the network of full connection layer generally are real numbers produced by a convex activation function to avoiding vanishing gradient problem while performing back-propagation steps. If an indication function,

$$sign(h) = \begin{cases} +1, & \text{if } h \geq 0; \\ -1, & \text{otherwise}, \end{cases} \tag{5}$$

is used to be the activation function, its gradient is zero for non-zero input. The process of back-propagation will be infeasible in training stage. For such a reason, the HashNet uses a hard tanh() function to be the activation function. The approximation of $sign$ (h) is based on the following equation.

$$sign(h) = \lim_{\beta \rightarrow \infty} \tanh(\beta h) \approx \begin{cases} +1, & \text{if } h \geq 0; \\ -1, & \text{otherwise}. \end{cases} \tag{6}$$

While evolving $\beta \rightarrow \infty, \tanh(\beta h)$ will approximate $sign(h)$. The initial value is $\beta = 1$, and the value of β will be increased in each converge of HashNet. By pre-training the initial β as the activation of the output layer, it enables easier converge of getting the binary hash codes of $(\boldsymbol{h}_i, \boldsymbol{h}_j)$.

4   Experimental Results

We compared the classifier-based approaches and the hash-based approach by testing the models on the image data set PASCAL VOC 2012 [16]. The total number of images in VOC 2012 data set is 11,540 including 5,717 images for training and 5,823 images for validation. Parts of the image set are multi-labeled. The image dataset are organized as two level concept hierarchy with 4 main concepts in the level 1 and 20 leaf concepts in the level 2, which are shown on detailed as follows:

Table 1. High-level concept retrieval for RF-1.

Concept	P	R	F1	AP
C0	0.886	0.346	0.498	0.551
C1	0.584	0.887	0.705	0.827
C2	0.436	0.830	0.572	0.775
C3	0.217	0.955	0.354	0.762

Table 2. High-level concept retrieval for RF-n.

Concept	P	R	F1	AP
C0	0.442	0.980	0.609	0.890
C1	0.601	0.967	0.741	0.938
C2	0.404	0.998	0.575	0.926
C3	0.968	0.287	0.443	0.798

Table 3. High-level concept retrieval for mRF.

Concept	P	R	F1	AP
C0	0.377	0.997	0.547	0.871
C1	0.448	0.992	0.617	0.947
C2	0.422	0.997	0.593	0.923
C3	0.355	0.981	0.521	0.802

Table 4. High-level concept retrieval for Hcode (30).

Concept	P	R	F1	AP
C0	0.968	0.817	0.886	0.942
C1	0.980	0.845	0.908	0.956
C2	0.927	0.723	0.812	0.914
C3	0.944	0.481	0.637	0.785

Person: person
Animal: bird, cat, cow, dog, horse, sheep
Vehicle: aeroplane, bicycle, boat, bus, car, motorbike, train
Indoor: bottle, chair, dining table, potted plant, tv/monitor, sofa

The evaluation of effectiveness for the validation set uses both unranked and ranked metrics. The unranked evaluation recognizes images with the top labeled concept. The basic measures contain precision (P), recall (R), F1 measure, and accuracy. The ranked evaluation counts the ranking order for the relevant concept as part of the measurement. The metrics of average precision (AP) and mean average precision (MAP) are used here.

The three classifier-based approaches RF-1, RF-n, and mRF are trained as the description in the Sect. 3.3. The RF-1 classifier returns the images according to their probability of the query leaf concept in descending order. The RF-n method will return the probability of each concept for an image and decide whether the image belongs to some concept or not. The mRF is the same as RF-1 except that there are different classifiers for the concepts of non-leaf levels. Images must be recognized from the top level. While upper-level concept being determined, the corresponding lower-level concept will be considered as the possible leaf concept of an image. However, if the upper-level was assigned a wrong concept, the lower-level concept must also be incorrect definitely. The hashing code method is marked as Hcode here. The concept of query image is determined by the top R similar images with minimal Hamming distance of the corresponding hashing binary codes.

There are four concepts in level 1 of VOC 2012 dataset which are denoted as C0, C1, C2, and C3, respectively. The retrieval results of level-1 concepts for the proposed models are shown in the tables of Tables 1, 2, 3 and 4. The metrics of P, R, and F1 are unranked. These measures concern the concept of top probability only. RF-1 distinguishes high-level concept through identifying the leaf-concept. Since the concept C0 is also one of the leaf concepts, the number of training examples is relatively large while training all leaf-concepts using a unique model. That is the reason that the precision of C0 dominates others in Table 1. On the contrary, RF-n trains individual model for each leaf concept. The precision of retrieval depends on the uniqueness of image features belonging to the corresponding leaf concept. The model mRF first picks the high-level concepts with high probability. The classifiers of leaf concepts then are used to make final decision. Such a retrieval strategy increases the recall, but its precision is lower. For hashing codes model, the precision and recall are relatively stable while the number of similar images is set to $R = 30$. As the results shown in the tables, the models RF-n, mRF, and Hcode have good ranked measure AP in general. The model Hcode is stable in the concept C0 which is also the leaf concept at the same time.

There are 20 concepts in level 2 of VOC 2012 dataset. The concept notation includes the high-level concept inside. For example, C1-1 is the child concept of C1. Since the number of concepts is large, only F1 measure is demonstrated as Fig. 3. For the concept of fine level, Hcode is superior to other classifier-based models. We note that the model mRF in fine level has much lower F1 measure than its results in level 1. The main reason is that the model mRF have to assign high-level concept first, and the fine level will be identified next. If the assignment of high-level concept is wrong, the fine-level is definitely incorrect. Therefore, the retrieval results of fine level are not so ideal while using the unranked measure concerning only the top rank images. As shown in Fig. 1, the model RF-n is relatively stable in comparison with the model RF-1. The model Hcode has better F1 measure for most of the concepts. But for the concepts C3-2 and C3-4, it nearly retrieves nothing correct. However, the ranked AP in Fig. 4 reflects that the models mRF have good performance although its F1 measure does not. The model RF-1 is always worse than other models. According to the three methods, RF-n, mRF, and Hcode, have advantage of their own. No one dominates the others in all leaf concepts.

We also test the number of major similar concept estimated by Hamming distance of binary codes. The results are given in Figs. 5 and 6. No heavy changing occurs obviously while the R value is between 20 and 50 in VOC 2012 dataset. Finally, the measure MAPs for the two concept levels are listed in Table 5. In summary, the model Hcode is the best one and the model RF-1 is the worst. The model RF-n is better than mRF in general.

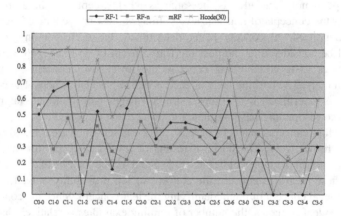

Fig. 3. F1 measure of leaf-level concept.

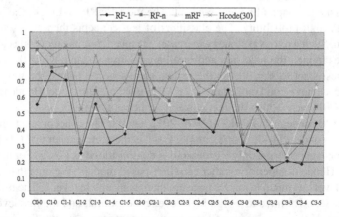

Fig. 4. AP values for leaf-level concept.

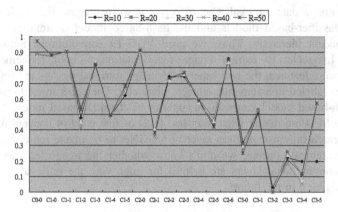

Fig. 5. F1 measure for different R values of Hcode.

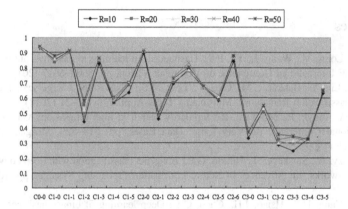

Fig. 6. AP measure for different R values of Hcode.

Table 5. MAP for 2-level concept hierarchy.

Level	RF-1	RF-n	mRF	Hcode (30)
High	0.729	0.888	0.886	0.899
Leaf	0.439	0.579	0.555	0.648

5 Conclusion

Retrieving images with similar semantic from a set of huge number of images is a challenge of this era. Due to the fast development of convolutional networks, image representation for a set of semantic concept can be learned effectively. Supervised deep hashing has been shown that the learned binary codes are effective and efficient in similar image retrieval. In the paper, we apply two supervised learning models for hierarchical image data retrieval. Our objective is to investigate the implicit characteristic of classifier-based models and the hash-based model on images labeled by semantic concept hierarchy. Thus, we used VGG-19 convolutional networks pretrained by ImageNet to extract image representation without refinement. After extracting image features, the two supervised image retrieval models are trained by a set of images labeled in a semantic concept hierarchy. The experimental results show that the hash-based model generally is superior to classifier-based model both in F1 measure and MAP no matter what in coarse level or fine level of concept hierarchy.

This research only uses the top layer of VGGNet to be image representation. For different hierarchical concept, we don't know whether the image features extracted from other convolutional layers are helpful for improving the retrieval results. It is worth investigating in the future.

Acknowledgments. This research was funded in part by Ministry of Science and Technology of Taiwan, R. O. C. under contract MOST 106-2221-E-024-017.

References

1. Cao, Z., Long, M., Wang, J., Yu, P.S.: Hashnet: deep learning to hash by continuation. In: The IEEE International Conference on Computer Vision, pp. 5608–5617 (2017)
2. Guo, Y., Liu, Y., Bakker, E.M., Guo, Y., Lew, M.S.: CNN-RNN: a large-scale hierarchical image classification framework. Multimedia Tools Appl. **77**(8), 1–21 (2018)
3. Gionis, A., Indyk, P., Motwani, R.: Similarity search in high dimensions via hashing. In: International Conference on Very Large Data Bases, pp. 518–529. Morgan Kaufmann, Edinburgh (1999)
4. Krizhevsky, A., Sutskever, I., Hinton, G.E.: Imagenet classification with deep convolutional neural networks. In: Advances in Neural Information Processing Systems, pp. 1097–1105 (2012)
5. Lin, K., Yang, H.F., Hsiao, J.H., Chen, C.S.: Deep learning of binary hash codes for fast image retrieval. In: The IEEE Conference on Computer Vision and Pattern Recognition Workshops, pp. 27–35 (2015)
6. Liu, H., Wang, R., Shan, S., Chen, X.: Deep supervised hashing for fast image retrieval. In: The IEEE Conference on Computer Vision and Pattern Recognition, pp. 2064–2072 (2016)
7. Simonyan, K., Zisserman, A.: Very deep convolutional networks for large-scale image recognition. arXiv preprint arXiv:1409.1556 (2014)
8. Szegedy, C., et al.: Going deeper with convolutions. In: The IEEE Conference on Computer Vision and Pattern Recognition, pp. 1–9 (2015)
9. Xia, R., Pan, Y., Lai, H., Liu, C., Yan, S.: Supervised hashing for image retrieval via image representation learning. In: Twenty-Eighth AAAI Conference on Artificial Intelligence (2014)
10. Wan, J., et al.: Deep learning for content-based image retrieval: a comprehensive study. In: 22nd ACM International Conference on Multimedia, pp. 157–166 (2014)
11. Wang, D., et al.: Supervised deep hashing for hierarchical labeled data. In: Thirty-Second AAAI Conference on Artificial Intelligence (2018)
12. Wang, H., Cai, Y., Zhang, Y., Pan, H., Lv, W., Han, H.: Deep learning for image retrieval: what works and what doesn't. In: IEEE International Conference on Data Mining Workshop, pp. 1576–1583 (2015)

13. Zhang, J., Peng, Y.: SSDH: semi-supervised deep hashing for large scale image retrieval. IEEE Trans. Circ. Syst. Video Tech. **29**(1), 212–225 (2019)
14. Zhe, X., Ou-Yang, L., Chen, S., Yan, H.: Semantic hierarchy preserving deep hashing for large-scale image retrieval. arXiv preprint arXiv:1901.11259 (2019)
15. Zhu, H., Long, M., Wang, J., Cao, Y.: Deep hashing network for efficient similarity retrieval. In: International Conference on Artificial Intelligence, pp. 2415–2421. AAAI Press, Phoenix (2016)
16. Visual Object Classes Challenge (2012). http://host.robots.ox.ac.uk/pascal/VOC/voc2012
17. Keras: The Python Deep Learning library. https://keras.io

Why Do People Back Crowdfunding Projects?

Ying-Feng Kuo[⊠], Cathy S. Lin, Chung-Hsien Wu,
and Tsung-Hsun Tsai

Department of Information Management, National University of Kaohsiung,
Kaohsiung, Taiwan, R.O.C.
frederickuo@gmail.com, miscathy@gmail.com,
jackiewu214@gmail.com, hero88214@gmail.com

Abstract. This study employs social cognitive theory as a theoretical foundation to empirically explore the influential antecedents of backing intention on crowdfunding platforms. We collected 221 valid samples from crowdfunding fan pages on Facebook in Taiwan, and the data were analyzed using the partial least squares (PLS) method. The results indicate that individual factors (i.e., backing self-efficacy, rewards, and empathy) and environmental factor (i.e., website service quality) all have a positive effect on behavior (i.e., backing intention). We conclude with a summary of theoretical implications and practical implications, along with an opportunity for IS researchers and practitioners to extend our knowledge of compelling future research possibilities.

Keywords: Crowdfunding · Social cognitive theory · Backing self-efficacy · Rewards · Empathy · Website service quality

1 Introduction

Crowdfunding is a practice of raising small amounts of money from a group of people on the Internet platform to back a specific project. Funders usually receive a product of the project or other forms of reward in exchange for the money pledged [15, 33, 48, 50, 65]. In crowdfunding mechanism, everyone can initiate a fund-raising proposal. Through rapid diffusion of ideas on the Internet, creators can directly seek financial support from Internet users and are no longer bound to use of traditional funding channels, such as investment by senior investors, bank loans or application for venture capitals. Compared to traditional funding, crowdfunding not only makes funding easier but also enables the public to quickly learn the goal and content of a project initiated by its creator. Besides, through the convenient online money collection channels, participants can solicit and collect donations from funders across the world. According to the recent report of 2017 Technavio market research, the global crowdfunding market is predicted to grow at an impressive compound annual growth rate (CAGR) around 17% during 2017–2021[1], noteworthy, the crowdfunding market is experiencing tremendous growth in Asia-Pacific area. With the growth of crowdfunding platforms and projects,

---

[1] http://www.businesswire.com/news/home/20170814005545/en/Global-Crowdfunding-Market—Segmentation-Forecast-Technavio.

© Springer Nature Singapore Pte Ltd. 2019
J. C.-W. Lin et al. (Eds.): MISNC 2019, CCIS 1131, pp. 14–28, 2019.
https://doi.org/10.1007/978-981-15-1758-7_2

the factors that affect funders' backing intention become an issue that deserves research and practical attention.

Social cognitive theory posits that individual factors, environmental factors, and behavior are reciprocally determined [4]; among the triangulate reciprocal mechanism, self-efficacy plays the core factor affects decisions about behavior [22]. Besides, prior studies of crowdfunding have found that funders' backing intention is motivated by rewards [33, 38, 42, 58]. In addition, we argue that personal empathy perception may influence one's support for a crowdfunding project. Collectively, this study considers backing self-efficacy, rewards, and empathy as individual factors to backing intention. Moreover, since crowdfunding is to raise funds on the Internet, the factors related to the online platform will have effects on the participants of crowdfunding project [1]. Previous studies have found that users' perception of website service quality is a determinant of their behavioral intention [41, 62]. Thus, we consider website service quality as the environmental factor of backing intention in this paper.

In sum, this paper adopts social cognitive theory as a foundation to investigate the triangulative relationships whether the above-mentioned individual factors (i.e., backing self-efficacy, rewards, and empathy) and the environmental factor (i.e., website service quality) affect funders' backing intention to support a crowdfunding project. This study not only enriches the existing literature by proposing a theoretical model that explains the funders' backing intentions on the crowdfunding platforms, but also extends the scope of application of social cognitive theory and confirm its applicability for explaining backing intention in crowdfunding. Based on the results, this study provides theoretical and practical implications and suggests directions for future research.

2   Literature Review and Hypotheses Development

2.1   Crowdfunding and Backing Intention

With the rapid development of the Internet and the growth of crowdsourcing, people can initiate a funding project online, allowing others to join the project by making donations or pledging money in exchange for a reward. To describe the relationship in a crowdfunding project, according to Agrawal et al. [1], those entrepreneurs, artists, and others who initiate projects or ventures are the role of "creators"; on the other hand, those investors, pre-buyers, and donors are under the label "funders." The difference characteristic is that creators can collect small amounts of money from large numbers of funders online without relying on the traditional funding approach [10, 15, 48, 50, 65]. It has been characterized as "potentially the most disruptive of all of the new models in finance"[2]. Once the collected fund reaches a threshold amount, project creators have the resources to fulfill the proclaim projects.

With a variety of forms of crowdfunding, which can be categorized into four models [15, 60]: (1) donation: funders do not receive any reward; (2) rewards: funders

---

[2] Goldman Sachs. 2015. "The Future of Finance - The Socialization of Finance, Part 3," retrieved September 7, 2017, from www.planet-fintech.com/file/167061/.

will be offered a product or a symbolic reward; (3) lending (also called peer-to-peer lending): funders can receive interest and capital repayments; (4) equity: funders can get shares with dividend or voting rights in return for their investment. In different forms of crowdfunding, funders play different roles depending on the distinct model and their motivations. For example, in lending- or equity-based crowdfunding, funders are like investors; they invest in projects they find worthy of support and expect a financial return on their investment. In donation-based crowdfunding, funders are viewed as donors, they give away money to support charitable or social welfare-seeking projects with no expectation of receiving anything in return. In reward-based crowdfunding, funders are more like consumers; they contribute small amounts of money to a project in return for a reward of a relative value [33, 38, 42, 50]. In Taiwan, most the existing crowdfunding platforms are implemented in a hybrid donation-based and reward-based model. Therefore, the focus of this study is placed on crowdfunding in a hybrid donation-based and reward-based model. In the hybrid donation-based and reward-based model of crowdfunding, funders may be motivated by the desire to obtain a product, therefore they play the role as consumers [33, 38, 42, 50].

Previous research of backing intention in crowdfunding has shown that motivations for backing a crowdfunding project include rewards [33, 38, 42, 58], simply giving a hand [33, 58], and being a member of the project (each project can be viewed as a temporary online community [16]) [33, 58]. Wash and Solomon [64] also revealed that some funders support a project because it feels good to do something right. Besides, project quality [38, 48] and creativity [23], creators' passion [38], and integrity of the team's information [13, 21] are also pivotal to the success of a project, because these elements affect funders' trust in the project. In reward-based crowdfunding, social influence also plays an important role [16, 42]. Based on the literature on consumers' purchase intention [11, 63], this study defines crowdfunding backing intention as "the strength of one's willingness to support projects on a crowdfunding platform".

2.2 Social Cognitive Theory

Base on social cognitive theory, we propose a research model to explore antecedents affecting backing intention on crowdfunding platforms. Social cognitive theory offers a sound theoretical background that has been extensively applied to explain individual behavior. This theory states that individual factors, environmental factors, and behavior are triangulating reciprocally determined [4]. In our research context, backing intention is the dependent behavior, while website service quality is the environmental factor. As to individual factors, the core concept of social cognitive theory surrounds self-efficacy (i.e., one's confidence in his/her capability to take a certain action) and outcome expectation (i.e., one's expectation of the result from a certain action) [3, 5]. In our research context, we will discuss one's belief about his/her capability confidence to support projects on a crowdfunding platform, and one's expectation of the outcome (rewards) of his/her support for the project. In addition, one's personal empathy perception and understanding for a crowdfunding project may also affect his/her support for the project. In summary, we propose backing self-efficacy, rewards, and empathy as individual factors that affect backing intention. In the following sections, we will review literature related to the research variables of backing self-efficacy, rewards, empathy, and website service quality and develop hypotheses.

Backing Self-efficacy
Self-efficacy refers to one's belief about his/her capability confidence to accomplish a given task [3, 5]. In other words, it is the degree to which one is confident that he/she can accomplish the given task. Based on the level of aggregation and stability over time and situations, self-efficacy beliefs can be classified into three types, including task-specific, domain, and general [34]. Task-specific self-efficacy is an individual's belief about his/her capability confidence to perform a specific task in a certain context. This type of self-efficacy is characterized by a very low level of aggregation and stability. Domain self-efficacy refers to one's belief about his/her capability confidence to perform tasks in a specific domain. For example, computer self-efficacy belongs to this type of self-efficacy [22]. General self-efficacy is a belief that one's capabilities confidence that can apply across different domains. This kind of belief is highly aggregated and stable. In crowdfunding, backing self-efficacy is a confident belief that one has the capability to back projects that he/she pays attention to. It is a kind of task-specific self-efficacy belief. To be more specific, backing self-efficacy is one's confidence in his/her capability to support a specific project on a crowdfunding platform.

Concerning the relationship between self-efficacy and behavior acts, self-efficacy is a determinant of intention and behavior. For example, computer self-efficacy has a positive effect on computer use [22]; self-efficacy for donating to the IRO is positively related to donation intention [17]; knowledge sharing efficacy affects the use of electronic knowledge repositories (EKRs) [40]; consumers' purchasing self-efficacy has a positive effect on their buying intention and behavior [53]; an individual's system self-efficacy is positively related to employees' extended use of enterprise systems [55]. Therefore, it can be inferred that people are willing to support a crowdfunding project when they believe they have the financial ability to support the project. Thus, we propose H1 as follows:

H1: Backing self-efficacy is positively related to backing intention.

Rewards
Outcome expectation is a judgment of the likely consequence of a certain behavior [3, 5]. Conceptually, it is similar to perceived benefit or perceived value. Several studies have pointed out that outcome expectation is an important construct that can be used to explain and predict individual behavior [39, 45]. Outcome expectations can be divided into three types, including physical, social, and self-evaluative [5]. Physical outcome expectations are experiences of pain or pleasure resulting from a behavior and the accompanying material losses and benefits. Social outcome expectations are the social approval or disapproval the behavior produces in one's interpersonal relationships. Self-evaluative outcome expectations refer to the individual's anticipated feelings about having performed a behavior [6]. According to the research of crowdfunding, most funders expect to receive a reward after offering their support for a project [33, 38, 42, 58]. Therefore, we classify rewards as a physical and self-evaluative type of outcome expectation.

Research evidences from various domains have suggested that reward is a determinant of behavior. For instance, when employees expect a reward for sharing knowledge, rewards will affect their attitude toward knowledge sharing [12]. In addition, rewards can induce a higher intention to participate in online communities [67], promote online interactions among users [37, 57], and encourage consumer involvement in product development to make more contributions or attract more consumers [30]. It is confirmed in many studies of crowdfunding that rewards have a positive effect on backing behavior [33, 38, 42, 58]. Therefore, we propose H2 as follows:

H2: Reward is positively related to backing intention.

Empathy

Empathy is a positive individual trait. It is one's perceptions to understand the motivations, values, and emotions of another [59]. Empathy can be viewed from cognitive and affective perspectives [47]. For the cognitive perspective, empathy is a cognitive process in which one assumes the role of another person and can understand and accurately predict the ideas, feelings, and behavior of another person [25]. The affective empathy perspective stresses the emotional or affective component of empathy [26], such as the emotional response to the perceived emotional experiences of others [61] or the response to another's emotional state [27]. Therefore, the importance of empathy lies in its perception to shorten the social distance between people and promote individual's social obligations [2].

Basil et al. [7] studied consumers' donations to charity and disaster relief and found that both empathy and self-efficacy are significantly positively related to donation intention. Batson [8] pointed out that empathy promotes helping behavior and prosocial behavior in individuals. Zhou et al. [70] also noted that empathy induces intentions to volunteer and donate and promotes tangible charitable behavior. The emergence of crowdfunding platforms allows both businesses and individuals an easier and faster way to fulfill their proposed ideas. Through the platforms, creators can effectively diffuse and express their creative ideas to win users' approval and further solicit their donations. Therefore, we infer that users' empathy perception can positively influence them to have the same emotional experience as the creator and further have a backing intention. Our third hypothesis is as follows:

H3: Empathy is positively related to backing intention.

Website Service Quality

The increasing prevalence of the Internet has facilitated the rapid growth of crowdfunding. As a bridge between creators and funders, crowdfunding platforms are becoming more and more important [69]. Crowdfunding platforms are designed for specific purposes, which differ to a certain extent from the purposes of general online shopping websites. Hence, how service quality plays a role in crowdfunding platforms should be an issue worth discussion. Service quality is the degree to which the service provided by the organization or a person meets or even transcends the service expected by customers [51]. Parasuraman et al. [52] developed SERVQUAL model to evaluate the service quality of the retail industry, which consists of five dimensions, including tangible, responsiveness, reliability, assurance, and empathy. As the online environment has become an increasingly common field for transactions, a number of studies have proposed their definitions and measurement scales for website service quality.

Website service quality refers to the degree to which the services provided by a website can effectively or sufficiently satisfy users' needs [28, 68]. The constituting dimensions of a website service quality scale may vary depending on the type of website to be evaluated. For instance, for information-based portal sites, Yang et al. [66] proposed to measure their service quality along usefulness of content, adequacy of information, usability, accessibility, and interaction. For e-commerce websites, Lee and Lin [44] proposed a scale consisting of website design, reliability, responsiveness, trust mechanisms, and personalization. Collier and Bienstock [20] identified functionality, information accuracy, design, privacy, and ease of use to measure website service quality. Bauer et al. [9] introduced a scale called eTransQual. This scale consists of functionality/design, enjoyment, process, reliability, and responsiveness. Udo et al. [62] proposed to measure a website's service quality by the ease of navigation, responsiveness, assurance, currency of information, and design quality. Kuo et al. [41] proposed four dimensions of website service quality, including content quality, navigation and visual design, management and customer service, and system reliability and connection quality.

Many studies have stated that website service quality strengthens user satisfaction [44, 66] and even influences users' behavioral intention [20, 28, 62] and purchase intention [44]. Literature related to crowdfunding suggests that the values of the crowdfunding platform have a positive effect on creators' behavior and funders' behavior [1]. In the research context, we infer that higher website service quality can lead individuals to have a higher backing intention. Thus, we propose H4 as follows:

H4: Website service quality is positively related to backing intention.

3   Research Methodology

3.1   Measurement Development

All measurement scale items were derived from previous studies and adapted to the crowdfunding context. Backing intention was measured with 4 items adapted from Bian and Forsythe [11] and Wang et al. [63]. These four items aimed at capturing the extent to which respondents would intend to back projects on a specific crowdfunding platform soon. For backing self-efficacy, three items adapted from Cheung and Chan [17] and Kankanhalli et al. [40] were used to capture the extent to which one's confidence in his/her capability to support a specific project on a crowdfunding platform. For rewards, three items adapted from Bock et al. [12] and Lee [43] were used to capture the extent to which funders expect to receive a tangible product or gift after offering their support for a project. For empathy perception, three items adapted from Basil et al. [7] were used to capture the extent to which funders have the same ideas and emotional experience as the creators. Website service quality is modeled as a second-order formative construct containing three first order reflective constructs (content quality, navigation quality, and interactivity quality). Content quality was measured with 3 items adapted from Collier and Bienstock [20], Kuo et al. [41], and Yang et al. [66]. Navigation quality was measured with 3 items adapted from Collier and Bienstock [20] and Kuo et al. [41]. Interactivity quality was also measured with 3 items adapted from Kuo et al. [41] and Yang et al. [66]. All the measurement items used a seven-point Likert scale, anchored from strongly disagree (1) to strongly agree (7).

As the respondents were native speakers of Mandarin Chinese and the questionnaire was originally developed in English, a translation and back-translation procedure were conducted. To ensure the adequacy and clarity of each question and identify potential problems in the questionnaire, a pretest was conducted using five crowdfunding experts and ten graduate students who had experienced in crowdfunding platforms such as Kickstarter, Indiegogo, flying V, and zeczec. The process of pretesting improved the content validity of the survey instrument.

3.2 Survey Administration

An online questionnaire was administered, and participants were limited to users who had experienced in browsing projects on any crowdfunding platforms. Before entering the formal survey page, each participant is required to answer if they have ever visited any crowdfunding platforms. If yes, the system would present the formal questionnaire. In this questionnaire, participants had to indicate which platform they most frequently use and answer questions based on such platform. If not, an thankful ending page will deliver to the participants because they do not qualify the requirement to fill out the survey.

Since the participants of this study must have had browsing experience with any crowdfunding platforms to seek the generalizability of the findings, this study recruited voluntary participants from crowdfunding fan pages on Facebook in Taiwan by employing chain-referral sampling. We also offered a coupon drawing as an incentive to increase participation willingness. The survey system was capable of recording cookies, access IP, and access time to detect repeated responses and invalid answers. The survey spanned four weeks and yielded a total of 221 valid responses.

In this sample, male respondents were the majority (55.2%). In terms of age distribution, respondents aged between 19–25 (38.0%) constituted the largest group, followed by those aged between 26–30 (26.2%). Most respondents had a college education level (52.5%), and those with a graduate school education level formed the second largest group (43.4%). Most respondents had a monthly income below NTD 20,000 (37.1%) or between NTD 20,000–50,000 (36.7%).

Further observing the sample demographic crowdfunding behavior, most respondents had a usage experience over 1 year (37.1%), followed by those with an experience between 4 months and 1 year (32.1%). This explains that most respondents in this survey had a certain level of experience with crowdfunding platforms. In terms of backing frequency, those who have never backed any project and those who have accounted for a similar proportion of the sample (51.1% vs. 48.9%). Among those who have backed a project before, the frequency of 4 times (15.8%) was dominant, followed by 2 times (11.3%). Finally, most of them had an average contribution amount between NTD 1001–5000 (21.2%), and those with an average contribution amount between NTD 500–1000 (19.9%) formed the second largest group. The above mentioned demographic statistics and crowdfunding funders' behavior are close to the investigation of Pew Research Center (2016)[3], illustrating the data sampling in this study approximates to the current crowdfunding population in the USA.

4 Results and Implications

4.1 Measurement Model Analysis

The reliability, convergent validity, and discriminant validity of the scale were tested using confirmatory factor analysis (CFA). The factor loadings for all the constructs ranged from 0.786 to 0.927 which were above the 0.70 guideline, indicating satisfactory item reliability for the measures [35]. Furthermore, the composite reliability (CR) values for the constructs ranged from 0.861 to 0.935 exceeded the recommended level of 0.70, indicating adequate internal consistency [14, 49]. The average variance extracted (AVE) values exceeded the suggested threshold value of 0.50, demonstrating the convergent validity of measures [29].

Discriminant validity was assessed using three criteria. First, when the loading of each measurement item on its assigned construct is higher than its loadings on all other constructs, and the cross-loading differences are much greater than the suggested threshold of 0.1 [32], the scales will be considered as having sufficient discriminant validity [18]. Second, the construct intercorrelation should be less than 0.71 to check the constructs have significantly less than half of their variance in common [46]. Third, the square root of the AVE of a construct should be larger than the correlations between the construct and all other constructs in the model [29]. The differences between loadings on assigned constructs and those on other constructs are greater than the threshold of 0.1. All correlations among constructs and the square root of the AVE are less than the 0.71 threshold. Further, comparing the square root of the AVE with correlations among the constructs indicates that each construct is more closely related to its own measures than to those of other constructs, which supports discriminant validity. Overall, the evidence of good reliability, convergent validity, and discriminant validity indicates the adequacy of the measurement model.

We further examined variance inflation factors (VIF) to assess the multicollinearity problem. A regression analysis that employed backing intention (BI) as the dependent variable and the other six variables as independent variables was performed. The VIF values of backing self-efficacy, rewards, empathy, content quality, navigation quality, and interactivity quality are 1.419, 1.375, 1.600, 1.998, 1.650, and 1.403 respectively, which are well below the suggested threshold of 3.3 [24]. Hence, the multicollinearity problem is not a concern for our data.

Common method bias (CMB) occurs when all data is self-reported and collected via the same questionnaire during the same period with cross-sectional research design. To test for CMB, we employed Harman's one-factor test [56]. The results showed that there was more than one factor and that the first factor accounted for 27.868% of the variance, lower than the 50% threshold value. After conducting Harman's one-factor

---

[3] http://www.pewinternet.org/2016/05/19/collaborative-crowdfunding-platforms/.

test, another test of CMB which compared correlations among the constructs was conducted following a procedure suggested by Pavlou et al. [54]. The highest correlation in the correlation matrix is $r = 0.575$, whereas evidence of common method bias ought to have brought about greatly high correlations ($r > 0.90$). Consequently, we thus conclude that CMB is not a major concern in our study.

4.2 Structural Model Analysis

The study employed the structural equation modeling (SEM) method to test the research model and hypotheses. When employing SEM, two types of methods, covariance-based techniques (CB-SEM) and variance based partial least squares (PLS-SEM), can be implemented. This study adopted a PLS-SEM approach because of the following advantages of this technique. First, PLS-SEM using component-based estimation, maximizing the variance explained in the dependent variable, does not require multivariate normality of the data, and is less demanding on sample size [18, 31]. Second, PLS-SEM is most suitable for models with formative constructs [19, 36], which is the case in this study. Third, whereas CB-SEM is regarded as being more appropriate for theory confirmation, but PLS-SEM is the preferred method for exploratory research, existed theory extension, and theory development [19, 31, 36]. This study is an exploratory research and the primary research objective is exploring the antecedents of backing intention in crowdfunding. Thus, PLS-SEM was appropriate for the current study. The SmartPLS 2.0 software package was used for our estimation. The bootstrapping procedure was implemented to provide reassurance that the results are not sample-specific by using repeated random samples drawn from the data. In this instance, the bootstrapping procedure was repeated until it reached 500 bootstrap samples [18].

We first tested the model with the entire sample consisting of 221 responses. To improve the internal validity, we also considered three other factors (gender, age, and education) as control variables in determining backing intention. The testing results support all the hypotheses (H1–H4). Backing self-efficacy has a significantly positive effect on backing intention ($\beta = 0.387$, $t = 5.583$), suggesting that funders with higher backing self-efficacy will be more willing to back projects on crowdfunding platforms. This finding echoes a similar finding of previous research that "consumers' purchasing self-efficacy is positively related to their purchase intention" [53]. Besides, the backing intention is significantly positively affected by rewards ($\beta = 0.097$, $t = 1.772$), indicating that providing tangible rewards can increase users' backing intention on crowdfunding platforms. This finding is consistent with evidence obtained by several qualitative observation-based studies [33, 38, 42, 50]. It is confirmed that empathy is positively related to backing intention ($\beta = 0.134$, $t = 2.195$), meaning that the better that funders understand the ideas behind a project and share the emotional experience of its creator, the higher the funders' backing intention. Results also indicate that website service quality has a significantly positive effect on backing intention ($\beta = 0.163$, $t = 2.166$). This suggests that funders are more willing to back a project when they perceive higher service quality on the website. The above four variables, including backing self-efficacy, rewards, empathy, and website service quality, explain 39% variance (R^2) in backing intention. With respect to the control variables, gender, age, and education level (coded as dummy variables) are incorporated into the model to test their effect on backing intention; we found that gender ($\beta = -0.007$, t = 0.098, p = 0.922), age ($\beta = 0.006$, t = 0.094, p = 0.925), and education ($\beta = -0.110$, t = 1.631, p = 0.104) do not have a significant effect on backing intention.

4.3 Theoretical Implications

Differing from previous crowdfunding related research, this study offers several important theoretical implications in the following ways. First, we adapt social cognitive theory as a foundation, which is a reciprocal determinism among person, environment, and behavior perspectives to explore the antecedents of backing intention on crowdfunding platforms. The current study offers a new theoretical understanding of the antecedents of funders' intentions on crowdfunding platforms and contributes to the literature by extending social cognitive theory into the context of crowdfunding.

Second, other than those qualitative studies observing the rewards for project funders (e.g., [33, 38, 42, 50]), this study empirically validates the significant effect of rewards by using a quantitative method as well as demonstrates rewards has a predictive relationship on backing intention.

Third, our results indicate that individual factors (i.e., backing self-efficacy, rewards, and empathy) and environmental factor (i.e., website service quality) all have effects on behavior (i.e., backing intention), explaining 39% variance in backing intention. Our results validate the finding of previous studies based on the qualitative observation that reward is one of the factors affecting backing intention [33, 38, 42, 50]. Moreover, the proposed model clearly shows the antecedents of backing intention and offers a new perspective for crowdfunding research. In this study, we extend the scope of application of social cognitive theory and confirm its applicability for explaining backing intention in crowdfunding.

4.4 Practical Implications

Can crowdfunding platforms really help raising funds for a project? This is probably the question that first comes to project creators' mind before they propose a project. When the project is launched, the copy-writing inside the projects is visible to funders considering whether to provide funds to support the project; therefore, diverse factors are aggregated around the project. Our results indicate that higher backing self-efficacy, higher perceived rewards, higher empathy, and higher website service quality can all lead to higher backing intention. The findings of this paper can facilitate understanding the operational strategies on crowdfunding platforms. Therefore, we propose the following suggestions for management of crowdfunding platforms and project creators.

1. Increasing funders' self-efficacy
Our results show that when funders have higher confidence for projects and in their ability to back projects on a crowdfunding platform, the more likely they are to back projects on the platform. The same finding can be obtained from both users with backing experience and users without. Our analysis further reveals that the predicting power of backing self-efficacy for backing intention is even greater among users

without backing experience. As to the suggested contribution amount, our survey results show that most respondents offer an average amount of NTD 1001–5000, followed by those giving away an amount between NTD 500–1000. The practical implication of this finding is that in order to increase users' confidence for the project and backing self-efficacy beliefs, creators can offer pledge options of a smaller amount to motivate those without backing experience to back the project. This strategy can turn more users without backing experience into actual funders.

2. Providing tangible rewards

Results show that when funders have a higher expectation of the tangible rewards for their donation, they will be more willing to offer their support. Besides, especially for the group without backing experience, tangible rewards have a greater appeal. Therefore, project creators are advised to offer funders a tangible product or any tangible reward in return. Tangible rewards can be a hand-written thank you letter, a gift or any product derived from the project. For the group with backing experience, how to take the incentive to the next level, from tangible rewards to intangible rewards, is an issue worth further discussion.

3. Inducing empathy in potential funders with compelling project descriptions

Potential funders show a higher backing intention when they have higher empathy for project ideas. In the group with backing experience, empathy has a significant effect on backing intention. Empathy can be a key factor that arouses continued backing intention in users with backing experience. Thus, when preparing a project's descriptions, project creators are advised to demonstrate their passions and articulate their creative ideas in a compelling manner. They should endeavor to influence potential funders to identify with them, approve of their ideas, and then financially support the project.

4. Improving website service quality

Potential funders' backing intention may increase with the degree to which the crowdfunding website's services effectively or sufficiently satisfy users' needs. Website service quality encompasses content quality, navigation quality, and interactivity. Based on these findings, we suggest that crowdfunding platform operators constantly update the website content, including projects or other related information. Besides, they should also focus on improvement of website navigation design and user experiences. The final aspect to improve is interactivity. For instance, crowdfunding platforms can provide a message board or a discussion forum to facilitate interactive discussions between funders and creators, and between funders. They can also be modified into a community-based website, allowing more people to propose their suggestions or questions and exchange opinions with others. This can benefit not just a single project but all the projects created on the platform. Crowdfunding platforms should not function simply as a platform for displaying projects. Instead, they should collect and utilize the collective power and wisdom from a larger group of people to make the raised funds more meaningful.

Acknowledgments. This research is supported by Ministry of Science and Technology (MOST 104-2410-H-390-021-MY2), Taiwan.

References

1. Agrawal, A.K., Catalini, C., Goldfarb, A.: Some simple economics of crowdfunding. In: Lerner, J., Stern, S. (eds.) Innovation Policy and the Economy, vol. 14, pp. 63–97. University of Chicago Press, Chicago (2014)
2. Aune, R.K., Basil, M.D.: A relational obligations approach to the foot-in-the-mouth effect. J. Appl. Soc. Psychol. **24**(6), 546–556 (1994)
3. Bandura, A.: Self-efficacy: toward a unifying theory of behavioral change. Psychol. Rev. **84**(2), 191–215 (1977)
4. Bandura, A.: Social Foundations of Thought and Action: A Social Cognitive Theory. Prentice-Hall, Englewood Cliffs (1986)
5. Bandura, A.: Self-efficacy: The Exercise of Control. Freeman, New York (1997)
6. Bandura, A.: Health promotion by social cognitive means. Health Educ. Behav. **31**(2), 143–164 (2004)
7. Basil, D.Z., Ridgway, N.M., Basil, M.D.: Guilt and giving: a process model of empathy and efficacy. Psychol. Mark. **25**(1), 1–23 (2008)
8. Batson, C.D.: Altruism and prosocial behavior. In: Gilbert, D.T., Fiske, S.T., Lindszey, G. (eds.) The Handbook of Social Psychology, vol. 2, pp. 282–316. McGraw-Hill, Boston (1998)
9. Bauer, H.H., Falk, T., Hammerschmidt, M.: eTransQual: a transaction process-based approach for capturing service quality in online shopping. J. Bus. Res. **59**(7), 866–875 (2006)
10. Beaulieu, T., Sarker, S., Sarker, S.: A conceptual framework for understanding crowdfunding. Commun. Assoc. Inf. Syst. **37**(1), 1–31 (2015)
11. Bian, Q., Forsythe, S.: Purchase intention for luxury brands: a cross cultural comparison. J. Bus. Res. **65**(10), 1443–1451 (2012)
12. Bock, G.W., Zmud, R.W., Kim, Y.G., Lee, J.N.: Behavioral intention formation in knowledge sharing: examining the roles of extrinsic motivators, social-psychological forces, and organizational climate. MIS Q. **29**(1), 87–111 (2005)
13. Boeuf, B., Darveau, J., Legoux, R.: Financing creativity: crowdfunding as a new approach for theatre projects. Int. J. Arts Manag. **16**(3), 33–48 (2014)
14. Bagozzi, R.P., Yi, Y.: On the evaluation of structural equation models. J. Acad. Mark. Sci. **16**(1), 74–94 (1988)
15. Bradford, C.S.: Crowdfunding and the federal securities laws. Columbia Bus. Law Rev. **2012**(1), 1–150 (2012)
16. Burtch, G., Ghose, A., Wattal, S.: Secret admirers: an empirical examination of information hiding and contribution dynamics in online crowdfunding. Inf. Syst. Res. **27**(3), 478–496 (2016)
17. Cheung, C.K., Chan, C.M.: Social-cognitive factors of donating money to charity, with special attention to an international relief organization. Eval. Program Plan. **23**(2), 241–253 (2000)
18. Chin, W.W.: The partial least squares approach for structural equation modeling. In: Marcoulides, G. (ed.) Modern Methods for Business Research, pp. 295–336. Lawrence Erlbaum Associates, Mahwah (1998)
19. Chin, W.W., Marcolin, B.L., Newsted, P.R.: A partial least squares latent variable modeling approach for measuring interaction effects: results from a Monte Carlo simulation study and an electronic-mail emotion/adoption study. Inf. Syst. Res. **14**(2), 189–217 (2003)
20. Collier, J.E., Bienstock, C.C.: Measuring service quality in e-retailing. J. Serv. Res. **8**(3), 260–275 (2006)

21. Colombo, M.G., Franzoni, C., Rossi-Lamastra, C.: Internal social capital and the attraction of early contributions in crowdfunding. Entrep. Theory Pract. **39**(1), 75–100 (2015)
22. Compeau, D.R., Higgins, C.A.: Computer self-efficacy: development of a measure and initial test. MIS Q. **19**(2), 189–211 (1995)
23. Davis, B.C., Hmieleski, K.M., Webb, J.W., Coombs, J.E.: Funders' positive affective reactions to entrepreneurs' crowdfunding pitches: the influence of perceived product creativity and entrepreneurial passion. J. Bus. Ventur. **32**(1), 90–106 (2017)
24. Diamantopoulos, A., Siguaw, J.A.: Formative versus reflective indicators in organizational measure development: a comparison and empirical illustration. Br. J. Manag. **17**(4), 263–282 (2006)
25. Dymond, R.F.: A scale for measurement of empathic ability. J. Consult. Psychol. **13**(2), 127–133 (1949)
26. Eisenberg, N.: Prosocial development: a multifaceted model. In: Kurtines, W., Gewirtz, J. (eds.) Moral Development, pp. 401–429. Allyn and Bacon, Boston (1995)
27. Eisenberg, N.: Emotion, regulation, and moral development. Ann. Rev. Psychol. **51**(1), 665–697 (2000)
28. Fassnacht, M., Köse, I.: Consequences of web-based service quality: uncovering a multi-faceted chain of effects. J. Interact. Mark. **21**(3), 35–54 (2007)
29. Fornell, C., Larcker, D.F.: Evaluating structural equation models with unobservable and measurement errors. J. Mark. Res. **18**(1), 39–50 (1981)
30. Füller, J.: Why consumers engage in virtual new product developments initiated by producers. Adv. Consum. Res. **33**(1), 639–646 (2006)
31. Gefen, D., Straub, D.W., Boudreau, M.C.: Structural equation modeling and regression: guidelines for research practice. Commun. Assoc. Inf. Syst. **4**(7), 1–70 (2000)
32. Gefen, D., Straub, D.W.: A practical guide to factorial validity using PLS-Graph: tutorial and annotated example. Commun. Assoc. Inf. Syst. **16**(1), 91–109 (2005)
33. Gerber, E.M., Hui, J.: Crowdfunding: motivations and deterrents for participation. ACM Trans. Comput.-Hum. Interact. **20**(6), 34:1–34:32 (2013)
34. Gibbons, D.E., Weingart, L.R.: Can I do it? Will I try? Personal efficacy, assigned goals, and performance norms as motivators of individual performance. J. Appl. Soc. Psychol. **31**(3), 624–648 (2001)
35. Hair, J.F., Anderson, R.E., Tatham, R.L., Black, W.C.: Multivariate Data Analysis. Prentice-Hill, Englewood Cliffs (1998)
36. Hair, J.F., Ringle, C.M., Sarstedt, M.: PLS-SEM: indeed a silver bullet. J. Mark. Theory Pract. **19**(2), 139–151 (2011)
37. Hennig-Thurau, T., Gwinner, K.P., Walsh, G., Gremler, D.D.: Electronic word-of-mouth via consumer-opinion platforms: what motivates consumers to articulate themselves on the Internet? J. Interact. Mark. **18**(1), 38–52 (2004)
38. Hobbs, J., Grigore, G., Molesworth, M.: Success in the management of crowdfunding projects in the creative industries. Internet Res. **26**(1), 146–166 (2016)
39. Hsu, M., Chiu, C.M.: Predicting electronic service continuance with a decomposed theory of planned behavior. Behav. Inf. Technol. **23**(5), 359–373 (2004)
40. Kankanhalli, A., Tan, B.C., Wei, K.K.: Contributing knowledge to electronic knowledge repositories: an empirical investigation. MIS Q. **29**(1), 113–143 (2005)
41. Kuo, Y.F., Wu, C.M., Deng, W.J.: The relationships among service quality, perceived value, customer satisfaction, and post-purchase intention in mobile value-added services. Comput. Hum. Behav. **25**(4), 887–896 (2009)
42. Kuppuswamy, V., Bayus, B.L.: Crowdfunding creative ideas: the dynamics of project backers in Kickstarter. UNC Kenan-Flagler Research Paper no. 2013-15 (2015). https://ssrn.com/abstract=2234765/. Accessed 15 Sept 2016

43. Lee, D.H.: The moderating effect of salesperson reward orientation on the relative effectiveness of alternative compensation plans. J. Bus. Res. **43**(2), 65–77 (1998)
44. Lee, G.G., Lin, H.F.: Customer perceptions of e-service quality in online shopping. Int. J. Retail Distrib. Manag. **33**(2), 161–176 (2005)
45. Lim, J.S., Noh, G.Y.: Effects of gain-versus loss-framed performance feedback on the use of fitness apps: mediating role of exercise self-efficacy and outcome expectations of exercise. Comput. Hum. Behav. **77**, 249–257 (2017)
46. MacKenzie, S.B., Podsakoff, M.P., Podsakoff, N.P.: Construct measurement and validation procedures in MIS and behavioral research: integrating new and existing techniques. MIS Q. **35**(2), 293–334 (2011)
47. Mehrabian, A., Epstein, N.: A measure of emotional empathy. J. Pers. **40**(4), 525–543 (1972)
48. Mollick, E.: The dynamics of crowdfunding: an exploratory study. J. Bus. Ventur. **29**(1), 1–16 (2014)
49. Nunnally, J.C.: Psychometric Theory. McGraw-Hill, New York (1978)
50. Ordanini, A., Miceli, L., Pizzetti, M., Parasuraman, A.: Crowd-funding: transforming customers into investors through innovative service platforms. J. Serv. Manag. **22**(4), 443–470 (2011)
51. Parasuraman, A., Zeithaml, V.A., Berry, L.L.: A conceptual model of service quality and its implications for future research. J. Mark. **49**(4), 41–50 (1985)
52. Parasuraman, A., Zeithaml, V.A., Berry, L.L.: SERVQUAL: a multiple-item scale for measuring consumer perceptions of service quality. J. Retail. **64**(1), 12–40 (1988)
53. Pavlou, P.A., Fygenson, M.: Understanding and predicting electronic commerce adoption: an extension of the theory of planned behavior. MIS Q. **30**(1), 115–143 (2006)
54. Pavlou, P.A., Liang, H., Xue, Y.: Understanding and mitigating uncertainty in online environments: a principal-agent perspective. MIS Q. **31**(1), 105–136 (2007)
55. Peng, Z., Sun, Y., Guo, X.: Antecedents of employees' extended use of enterprise systems: an integrative view of person, environment, and technology. Int. J. Inf. Manag. **39**, 104–120 (2018)
56. Podsakoff, P.M., MacKenzie, S.B., Lee, J.Y., Podsakoff, N.P.: Common method biases in behavioral research: a critical review of the literature and recommended remedies. J. Appl. Psychol. **88**(5), 879–903 (2003)
57. Rafaeli, S., Raban, D.R., Ravid, G.: How social motivation enhances economic activity and incentives in the Google Answers knowledge sharing market. Int. J. Knowl. Learn. **3**(1), 1–11 (2007)
58. Ryu, S., Kim, Y.G.: A typology of crowdfunding sponsors: birds of a feather flock together? Electron. Commer. Res. Appl. **16**, 43–54 (2016)
59. Salovey, P., Mayer, J.D.: Emotional intelligence. Imagin. Cogn. Personal. **9**(3), 185–211 (1990)
60. Simons, A., Weinmann, M., Tietz, M., vom Brocke, J.: Which reward should I choose? Preliminary evidence for the middle-option bias in reward-based crowdfunding. In: Proceedings of the 50th Hawaii International Conference on System Sciences, Big Island, HI (2017)
61. Stotland, E.: Exploratory investigations of empathy. In: Berkowitz, L. (ed.) Advances in Experimental Social Psychology, vol. 4, pp. 271–314. Academic Press, New York (1969)
62. Udo, G.J., Bagchi, K.K., Kirs, P.J.: An assessment of customers' e-service quality perception, satisfaction and intention. Int. J. Inf. Manag. **30**(6), 481–492 (2010)
63. Wang, Y.S., Yeh, C.H., Liao, Y.W.: What drives purchase intention in the context of online content services? The moderating role of ethical self-efficacy for online piracy. Int. J. Inf. Manag. **33**(1), 199–208 (2013)

64. Wash, R., Solomon, J.: Coordinating donors on crowdfunding websites. In: Proceedings of the 17th ACM Conference on Computer Supported Cooperative Work and Social Computing, Baltimore, MD (2014)

65. Wheat, R.E., Wang, Y., Byrnes, J.E., Ranganathan, J.: Raising money for scientific research throughout crowdfunding. Trends Ecol. Evol. **28**(2), 71–72 (2013)

66. Yang, Z., Cai, S., Zhou, Z., Zhou, N.: Development and validation of an instrument to measure user perceived service quality of information presenting web portals. Inf. Manag. **42**(4), 575–589 (2005)

67. Yen, H.R., Hsu, S.H.Y., Huang, C.Y.: Good soldiers on the web: understanding the drivers of participation in online communities of consumption. Int. J. Electron. Commer. **15**(4), 89–120 (2011)

68. Zeithaml, V.A., Parasuraman, A., Malhotra, A.: Service quality delivery through web sites: a critical review of extant knowledge. J. Acad. Mark. Sci. **30**(4), 362–375 (2002)

69. Zheng, H., Li, D., Wu, J., Xu, Y.: The role of multidimensional social capital in crowdfunding: a comparative study in China and US. Inf. Manag. **51**(4), 488–496 (2014)

70. Zhou, X., Wildschut, T., Sedikides, C., Shi, K., Feng, C.: Nostalgia: the gift that keeps on giving. J. Consum. Res. **39**(1), 39–50 (2012)

DHUM: A Dynamic Human Urban Mobility Model for Smart Public Transport

M. Saravanan$^{(\boxtimes)}$ and Perepu Satheesh Kumar

Ericsson Research, Ericsson India Global Services Pvt. Ltd., Chennai, India
{m.saravanan,perepu.satheesh.kumar}@ericsson.com

Abstract. Smart transportation helps people in the smart city to reach their destinations securely on time with the minimum waiting time. In general, the public transport system is scheduled based on different time of the day which not depends on other parameters like traffic, dynamic way of an understanding number of traveling passengers, etc. Due to this, some of the routes are heavily crowded and some are not. These can result in wastage of resources and not bring any smartness to public transport and it also causes discomfort to the citizen's. In addition, this may lead to higher waiting times and force people to use their personal vehicles, ultimately, it increases the pollution level in cities. In this work, we propose a new method to schedule public transport by predicting the traveling passengers in each location. For this, we explore a way to understand the mobility patterns of citizen's. *DHUM,* a dynamic human urban mobility model will be proposed for effective urban planning and transportation by exploring the use of mobile crowdsourcing data and heat maps in addition to their mobility pattern. This can help authorities to schedule public transport in real-time intelligently so that the citizen's hassles may be governed to make them reach their destination on time. With the use of Dakar city data, we have rationalized that our proposed method can result in a decrease in pollution and subsequently it helps to increase the people's comfort in future smart cities.

Keywords: Call Detail Record (CDR) · High-density zone · LatLong · Geographic Information System (GIS) · Human mobility model

1 Introduction

More than half of the world's population now lives in urban areas. Due to this enormous number of people move to urban areas created a new set of problems such as overcrowded in public transport, pollution, waste management, water scarcity, etc., [1]. An upsurge in population and urbanization around the world are causing increasing problems for the movement of the citizens in their city. To handle these problems, we need an intelligent way to manage the resources available. In this paper, we considered the case of managing public transport in an optimal manner such that the pollution generated from the city will be under control and it increases the comfort level of citizens. This one way can also help in addressing global warming.

Nowadays, we witness all over the world that most people prefer to travel on their own vehicles rather than public transport. The reasons for this is due to crowded public

© Springer Nature Singapore Pte Ltd. 2019
J. C.-W. Lin et al. (Eds.): MISNC 2019, CCIS 1131, pp. 29–42, 2019.
https://doi.org/10.1007/978-981-15-1758-7_3

transport or less comfort in travel. Traveling in their own vehicles can generate more pollution in the city and which will slowly cause global warming of the planet. In addition, this can create too many traffic chaos in the city. A solution for this problem is to make public transport available for the user and provide enough comfort to the public who intend to use.

Recently in cities, one can witness plenty of mobility challenges which are not limited just to traffic congestion. They are also about efficiently connecting (time, cost, effort) different neighborhoods with public means of transport, helping citizens and professionals at the last mile journey, giving access to the critical stations (train, airport, buses) with multiple means and from multiple regions, offering a variety of options to the people to move around (including bicycle), offering of parking slots, and many more. It is also about understanding how citizens move every day for city officials to plan accordingly the location of stations, bike routes, and traffic lights, as well as to optimize the schedule of each city activity without disturbing others. With the use of IoT and AI-enabled solutions, cities can be improved and solve or at least reduce some of the main urban transportation issues.

Most of the cities all over the world, public transportation is scheduled based on the timings of the day rather than the number of people traveling and by considering city traffic. For example, assume there is a work area, where most people will travel in the morning and back home in the evening [11]. One expects more buses to fly on the route in the morning and evening. Suppose, assume the transport is scheduled for every 15 min in the morning. Also, assume there are 100 people waiting for transport at a given time. In this case, the transport (here, in this case, it is a bus) will be too crowded and it can result in discomfort to the passengers. On the other hand, if we can schedule transport based on the crowd not on the time basis, it can help the passengers to be more comfort i.e. by scheduling two or three buses on the same time to avoid over-crowding in one single bus.

However, this is not an easy task as we cannot schedule transport on the fly due to limited resources. In general, to schedule any bus it will take at least two hours for the agency who need to handle transport operation in the city. Hence, in this work, we chose to predict the crowd in each area and their mobility pattern two hours in advance to enable online scheduling of transport which can generate a good comfort level for the passengers. In this paper, we propose an extrapolative model named as *DHUM* (Dynamic Human Mobility Model) to predict the mobility patterns of the users using the data collected from social media accounts and/or wearable devices like mobile phones, smart watches, etc. It enables two types of scheduling methodologies to handle the situation. Predict the mobility patterns of the office goers daily and schedule public transportation a day in advance. For this, we use Call Detail Records (CDR's) of persons in the city. We will consider this as a global model in this paper. The social media information of the users can be detected and use that information to re-schedule the public transport in near real-time i.e. two hours in advance, will be considered as a local model. With the combination of both these models, our proposed system can predict the exact movement of citizens on the location.

In the "Internet Era", everything is digitally computed and the solution to each problem lies in the modern-day systems like mobile phones, laptops, and other pervasive devices. True to their name, these pervasive devices are present in every item used by human beings. These can be integrated with day to day articles like watches, cars, refrigerators, washing machine, etc. Pervasive devices can also be embedded into the human body in the form of a chip that keeps track of human mobility. These devices not only make human life easier but also help to manage an entire city even of metropolitan scale. The mobile service providers apart from maintaining the customer records also help in collecting "digital census" which acts as a means of keeping a count on the population.

DHUM is built to serve the purpose of managing an effective urban plan for a region along with a proper transportation schedule that connects the metropolitan city to the outskirts and the developing cities nearby. This is done by using the concept of crowdsourcing. The model operates with the help of dynamic data visualization that mimics an accurate movement of humans. With the help of this, urban planning and optimized transportation plans can be built for a city [7, 9]. A machine learning model is used that calculated the probable population of that region in the future, say after five years. To do so, it works on the trends found in the past and tries to take it to the future. So, with the help of the projected population, the required land resources can be identified to make a proper urban plan. Not only at calculating mobility but, a visual representation is also provided with the help of Global Information Systems that help in plotting the mobility of people using various visualization tools. A feedback loop is used that automatically updates the system so that the predictions made are highly accurate. Our main contributions in this paper are

- To predict the mobility pattern of persons in the city using the proposed DHUM model.
- To come with a method to schedule public transport to address user mobility and increase their comfort level of local citizen's.
- To alter the predictions in near real time to improve the efficiency of scheduling.

This paper is organized as follows. Section 2 discusses the related works in terms of mobility prediction and scheduling mechanism. In Sect. 3, we discussed the proposed model to predict mobility pattern and to do effective scheduling. Results of the proposed method on real-time CDR data of Dakar city are presented in Sect. 4. In Sect. 5, we discuss the conclusions and future recommendations.

2 Literature Review

The current pervasive models include various input/output devices, handheld devices or embedded chips that can be used concurrently with other applications. In the mobile and networking field, it can be used for connecting multiple devices with one another which apart from helping in information exchange, can also be used for statistical analysis of human mobility.

The existing system for human mobility model is an improvement when compared to the age-old procedure of relying on the physical census data [3–5]. The existing system relies on what is called as "Call Detail Records (CDR)" [1]. These records are generated for each registered mobile number. It contains the details of the time of origination of call, the duration of the call, the tower details of the origination and destination of the call [2]. Not just calls, the CDRs also store the information regarding the details of messages sent from one mobile to another. The CDRs are generally generated and stored for billing purposes [2]. They, however, can also be used to track human mobility. The spatiotemporal information on human mobility can be analyzed with the help of these CDR's.

This model of human mobility represents with CDRs can thus capture the human movements in a digital form which is also more accurate when compared to the previous generation's mobility model. But, there are some common problems that can be identified with this system. They are:

1. Only the geographical information of the tower is registered as the location (that is, the latitude and longitude coordinates). From this, it will be difficult to say if that location corresponds to home or work or any other place.
2. The spatiotemporal information is not highly accurate. That is, the spatial location only depicts the cellular towers and not the actual location of the person and the temporal information is registered only when the mobile phone is put to use. In other scenarios, the mobility becomes static, which is not the case.

Many approaches were proposed to build a proper human mobility system, out of which few uses synthetic CDR and the probability models to predict the actual location of the person [1]. Data mining techniques can be applied to extract the location of a person. These models are equally effective where one model does an operation which is original and not tried before. An ideal system for human mobility is one which combines all these aspects into a single model and simulates real-time mobility that nearly replicates the movement of people.

Thus, it can be inferred that the models built to depend on the applications of the mobility pattern [1, 2, 10]. That is, a different approach must be used for statistical analysis that solely concentrates on taking the census. This need not portray that much of visualization. Many other mobility models can also be built for the purposes of targeting people who spend most of their time in their home so that an effective forum for crowdsourcing can be built. This model must have the feature of high connectivity that connects the home location and the work locations so that the work can be effectively distributed to enable crowdsourcing.

There is a considerable amount of literature on the scheduling of public transport. In [14], the authors used ARIMAX model to predict the optimally schedule public transport. The problem with this method is based on the history of the number of people who travel by buses every day. In another work [16], the authors presented a method to schedule the public transport such that the waiting time of people traveling is minimum. As with the previous method, this is also based on the history of the data available. Other types of work which discuss scheduling of the public transport based on the demand [15, 16]. However, the problem is that the demand is predicted by asking the user. In this work, we propose to predict the demand in a location by the proposed DHUM model which utilizes the CDRs of the persons. This will be the most accurate prediction as everyone have their own mobile phones.

The model that we are proposing in this paper uses data visualization tools like Gephi, Google Earth, Google Charts, High charts to provide a dynamic visualization for human mobility. A timescale is provided with each mode of visualization so that the actual mobility can be provided in relation to time. Thus, an accurate spatiotemporal model is built which is highly dynamic and adaptive to human mobility in addition to crowdsourcing and heat maps inputs.

3   Proposed Work

As discussed in earlier sections, the main contribution of the paper is to schedule the public transport based on the identification of exact locations of the person movement. For this, we use the CDR's of related persons in the city along with the social media feeds. We propose to identify human mobility with the help of DHUM, a new dynamic human mobility model in this paper.

The human mobility model that we have projected in this paper is highly dynamic in nature, effective and time-saving. With the help of the high level of data visualization provided by DHUM, one can easily identify the sparsely populated areas and densely populated areas. In general, most of the people who are working on day shifts will face the problem of congestion and traffic. Hence, from the mobility pattern at night, it is easy to identify the locations of people and we classify them as their home locations. In addition, we observe the mobility pattern at morning and we name them as their office locations. In this case, the mobility pattern of users will be observed in the morning and their return in the night. However, it should be noted that the time of arrival and departure of all these persons is not the same. The challenge is to predict the exact time of departure and arrival so that the public transport can be scheduled on time for the same. As discussed in the introduction, the proposed model uses both the call data records of the customers along with the social media updates. Now, first, we discuss the construction of mobility patterns from the CDRs of the customers for building a local model.

3.1   Working Model of DHUM

Over the timescale, the areas which are always densely populated can be identified as residential areas and from the mobility information, the areas which are visited every day in a regular pattern can be observed as the work areas. Using predictive modeling, it can be analyzed that, with the increase in population and the urbanization of metropolitan cities, it can be analyzed that the work areas are expanded to the outskirts of the cities which must be targeted for urban planning and transportation.

This model, after identifying the densely populated locations, the home and work locations can do an effective transportation planning that decides both an optimal route to the work location and the number of vehicles required in each route depending upon the density.

Figure 1 illustrates a very simplified version of how the DHUM works. The various components are interconnected and a feedback loop is used to improve the system to make it more accurate. The feed is given in an automated fashion which compares the actual with the suggested transportation model. These components are explained in detail:

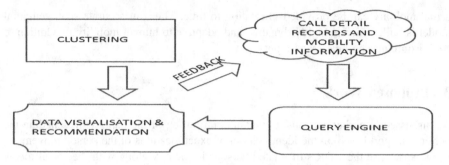

Fig. 1. Block diagram of DHUM

1. Clustering Analysis

In this paper, we propose to cluster the CDR data in the city to ten different clusters as given in Table 1. The value ten considered here is an example and the user can choose any value depending on the area of the city. In order to arrive at these ten segments, the k-means clustering model was applied. These points hold the details of the aggregated CDRs from which mobility pattern can be applied. The center of each location represents the common pickup points of the people in the location. Thus, a fuel effective transport system can be built.

Here, the cluster that is highly huddled can be identified as the high-density zone and can be effectively split up. Such high-density zones must be given top priority when it comes to transportation planning. This, in turn, will attract more people towards that region thereby making it denser. Thus, the process of migration towards a location, the effectiveness of urban planning and transportation planning are interdependent. Therefore, urban planning in these areas must be carried out with utmost care.

Table 1. K-means clustering method

***Input:* -** *k: The number of clusters the data should be divided*
1:   Partition the Data Set into k-nonempty subsets
2:   Compute seed points as the centroids of the cluster of the current partition. The centroid is the center of the cluster.
3:   Assign each object of the cluster to the nearest seed point.
4:   Go back to step 2. Stop when no more assignment is possible.

Mobility mapping is the process of mapping the Latitude and Longitude (LL) information provided in the CDR with the geographical region. The actual site information and the calls per site (mobility) can be illustrated with the help of mobility mapping. This shows the distribution of the location of people. In this paper, we have carried out the process of mobility mapping in three levels of abstraction. They are:

- *Lowest Level of Abstraction:* The bottom-most level of mobility abstraction gives a picture of the mobility as such. That is, the actual physical towers are represented and the movement is shown. To do this, we have used Google Earth so that the LL information can be accurately plotted in the actual geographic location. This model, even though has a high level of accuracy when it comes to displaying mobility, will be hard to use as it becomes difficult to infer pattern when the size of mobile information grows.

- *Mid Level of Abstraction:* The next level of abstraction is made after the implementation of K-Means algorithm. Thus, here, instead of taking the LL position of all the towers, they are clubbed together to form ten distinct points. The mobility information from and to these points are then depicted with the help of the visualization tool Gephi [8]. This mode of visualization does not provide the 'in-depth-visualization' that was provided in google earth, but statistical information can be gathered from this. Since this also uses the LL information of the towers, the absolute position of the towers can be identified. Thus, this level provides a little bit of both visualizations as well as statistical analysis.
- *Top Level of Abstraction:* The top-most level of data abstraction uses Google Charts for representation. This concentrates solely on the relation between time and mobility information. Thus, the LL information is not used here and only relative positions are used for analysis. Once this is done, the location, say region 1 can be identified as the High-density-zone depending upon the mobility pattern. Thus, this level can be used to provide a high level of urban planning. Once the density of movement is analyzed, the LL information is imposed upon the data to do further transportation plan between the points.

One common feature that can be identified with all the three levels of abstractions is that they are all dynamic in nature and they show the change in the mobility pattern with reference to a starting point. All these three levels use timescales to help visualize the change in mobility. This timescale can be altered to show the mobility changes at different intervals. For example, the entire data can be rolled up to show the monthly change in mobility or can be highly drilled-down to map the mobility changes from every minute. In this paper, we have used an aggregated monthly analysis for Google chart, daily analysis for gephi and every minute changes for Google earth.

Another advantage of mobility mapping is that we can visually spot the outliers easily. That is, if there is any region with no mobility, it can be clubbed to a near-by dense location to make it easily accessible. In a similar manner, a vastly dense location can be split up into zones for ease of transportation and traffic control. In this way, we can predict the mobility changes for different time scales ranging from every minute to every month.

2. Predictive Analytics and Recommendation Model
This block helps in predicting the mobility model for the future with the help of the current mobility pattern analysis. Thus, appropriate resource allocation can be done for the future. The concept of supervised learning is used to predict the class label of a new instant. Thus, with the help of this block, the population density can be predicted in advance. Also, the exact urbanization of the outskirts can be predicted. This will help in detailed urban planning which includes the expansion of IT Parks, shopping malls and boosts the real estate business.

A recommendation/prediction matrix can be built to accurately compare a similar trend observed in the past. The items can be placed along the x-axis. The items here correspond to the regions that are separated with respect to LL Coordinates. Along the y-axis, we can have the predicted value of population growth or the population density of a region. When similar trends are observed, the regions can be grouped together and a common urban and transportation planning system can be applied.

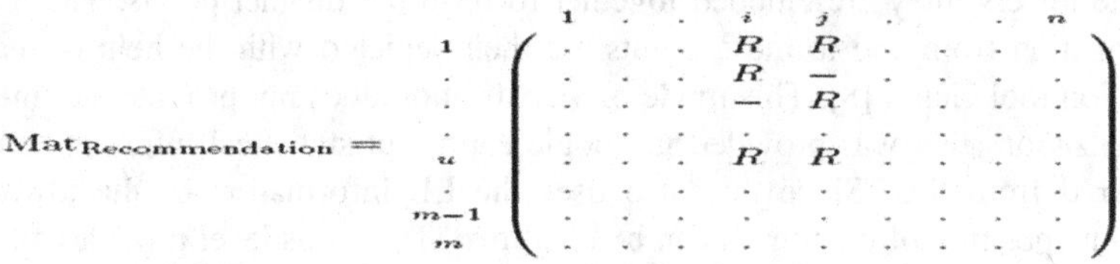

$$
\text{Mat}_{\text{Recommendation}} = \begin{matrix} & & & & i & j & & & n \\ 1 & & & & R & R & & & \\ & & & & R & - & & & \\ & & & & - & R & & & \\ u & & & & R & R & & & \\ m-1 & & & & & & & & \\ m & & & & & & & & \end{matrix}
$$

Fig. 2. Recommendation matrix

Using the recommendation matrix shown in Fig. 2, a similarity vector can be built between two sites i and j. This similarity is given with the help of Pearson (Correlation) – based Similarity formula

$$
sim(i,j) = \frac{\sum_{u \in U} \left(R_{u,i} - \bar{R}_i \right) \left(R_{u,j} - \bar{R}_j \right)}{\sqrt{\sum_{u \in U} \left(R_{u,i} - \bar{R}_i \right)^2} \sqrt{\sum_{u \in U} \left(R_{u,j} - \bar{R}_j \right)^2}}
$$

The value of $sim(i, j)$ has a value that always lies between -1 and $+1$. When it is -1, it means that the items have a negative correlation and are not related. When the value is $+1$, then it means that they have a positive correlation and they show similar trends (population density, transportation plan).

3. Feedback Loop

The feedback loop is included in the system to improve its efficiency. The feedback mechanism is explained as follows: If the prediction is made and it matches with the actual pattern then it triggers positive feedback and the relevant learning model is applied for the future cases. If the prediction is made which deviates from the original value, then it throws negative feedback so that the concerned learning algorithm is modified to ensure that the system keeps up with the pattern followed in the region.

4. Query Engine

The query engine is used to query the mobility pattern between locations in the city. The input to the engine will be providing the source and destination locations in the city. The output of the query engine contains mobility patterns between the location against the time of the day. At the end of this step, we obtain the mobility pattern of the customers and let us call it as a global model. In the next section, we describe the schedule of public transport based on the predicted mobility pattern.

3.2   Scheduling of Public Transport

As discussed at the end of the introduction, we schedule the public transport in two steps (i) Construction of the global model from CDR data records and (ii) Local model to update the predictions. The block diagram of this part is shown in Fig. 3. We discuss the model in detail in subsequent sections.

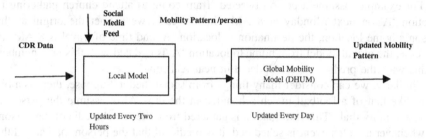

Fig. 3. Block diagram of the proposed scheduling step

i. Scheduling in a day in advance: From Sect. 3.1 of the work, we obtain the mobility patterns between ten different locations in the city. It is assumed that these ten locations have 10 different hubs. From the mobility patterns, we can understand the persons' departure time at location A and person arrival time at location B. In this way, we obtain data for all the persons in the city, who travel frequently from one location to another location. Based on these patterns we can predict the number of people arriving at the location traveling to another location at any given time. This is useful so that we can arrange public transport ahead of time. Usually, this mobility pattern won't change in a day and hence we propose to run this model per day to schedule the bus transport. For scheduling the public transport, we use the CDR's of persons in each area as it will give the mobility pattern of all the people.

ii. Scheduling in near real-time: From the previous section, we obtain the schedule of bus transport per day. However, since the mobility pattern is not updated in near real-time it can lead to false results. Hence, it should update in near real-time for efficient scheduling. For this, we considered the social media of the persons in the given location for near real-time scheduling.

From the social media feed of the persons, we extract important events, events in the city, interests, etc. Once, this information is extracted we able to extract easily the mobility pattern of the passengers. For example, assume the person updated as going for a vacation for seven days. In addition, assume there are several persons like that. In this case, if we schedule bus transport without considering the situation, it will result in wastage of resources. Hence, in this case, we remove the count of the people from predictions in the previous section and schedule the buses for the new predictions. In short, the model constructed from the previous section is a global model. In this section, we update the predictions based on the social media feed.

For updating the prediction model, we use the social media feed of the persons in the city. To do this, we extract the full name of the subscriber present in the CDR data records and crawl the twitter feeds of the person. These tweets are collected per person and stored in the database. From the tweets, we should extract the features such as the original destination of the person, time of travel, accompanying people, etc. In this paper, we extract these minimum features only. However, it is possible to extract additional features also which is outside the scope of this paper.

For example, assume a person tweeted 'I am going to attend church gathering in location 'A' on next Monday at 8 AM'. In this case, we extract the origin as the person's home location, the destination is location 'A' and time of travel is 8 AM. In this case, the global model from home to location 'A' is updated with this new number. In this way, the predictions are updated in near real time.

Similarly, we can consider many things from social media. Suppose, there is a big event like that of a football match scheduled in the city. Also, assume the person is interested in football. This information is gathered from the social media of the person. So whenever a match event is scheduled, it is predicted that the person may attend the football match with high probability and hence this count can be decreased from the predictions of a global model. We collect similar persons like this and we decrease the count from the predictions of the global model so that transportation can be scheduled based on the update of crowdsourcing at any given time. As mentioned in the introduction, this cannot happen in real time as scheduling of public transport requires at least two hours due to logistic constraints. Hence, we propose to update predictions for every two hours so that schedule is made. Hence, this scheduling is not in real time but it is in near real time. In summary, the global model is updated daily based on the call data records of the customer. The global model is used to predict the number of people traveling from location A to location B at any given time. The predictions from the global model are not always correct as we are not considering instantaneous updates from the people in the locality. For this, we consider the social media of the person to get instantaneous updates from the user and use them to update the predictions from the global model. This is run for every two hours to ensure the public transportation is scheduled optimally.

4 Experimental Results

4.1 Global Model

As discussed, we construct the global model from the CDR dataset of the customers. The CDRs are obtained from Dakar city citizens, the capital of Senegal in West Africa. Those CDRs give the information like the unique subscriber ID, the telecom plan that the subscriber has chosen, the number of calls made, the sender and recipient of messages, the LL information of the tower where the calls were registered, the time when calls were made or messages were sent and finally, the duration of the calls. The details collected from the Dakar region corresponds to the urban details. The objective here is to predict the mobility pattern between locations based on CDR data.

Simple SQL queries are used to group CDRs of different site details of the same region. Anonymization is ensured in the system by removing (encryption can also be used) the unique subscriber ID. This is done so that only the relationship between the region and mobility is got and the privacy of the subscriber is maintained. The subscriber information could be used only by the appropriate Mobile Service Providers who would wish to make a personalized recommendation to the subscribers depending upon their unique mobility pattern and their call information.

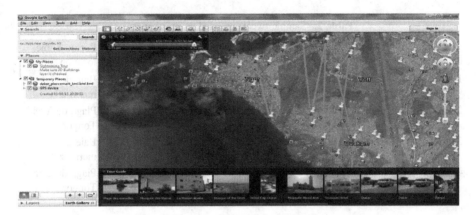

Fig. 4. Mobility representation in Google Earth

We divided the Dakar into ten different clusters and identify the person in each cluster. In this exercise, we found that each region has 20K subscribers approximately. Their mobility pattern is mapped and it is identified that most of the people in the Dakar are employees and travel for their work daily. From this, we build the model which will predict the locations of the person every 15 min. A sample mobility pattern obtained at one location for the 15-min interval is shown in Fig. 4. This is to ensure that public transport is scheduled at the given time and given a location. In the next section, we discuss them in detail.

4.2 Scheduling of the Bus Transport

In this section, we discuss the results of the proposed method to schedule the public transport. As discussed, we obtain the mobility patterns of the customers from their CDR data records. After we obtain these records, we schedule the bus transport at the time and location as we predicted. For an example, when we predict at 9:00 AM, 100 people travel from location Biscuiterie to Pikine. In this case, we require at least two buses in this route so that the buses are not crowded. In normal case, the buses in this route are schedule for every 15 min in the morning [13]. In this case, if we use the proposed method we can schedule two buses at this time thereby giving more comfort to the passengers and also motivating them to take public transport.

This is the global model for us. In next, we discuss the problem of updating the predictions of the global model. For this, we collect the social media feed of the persons staying in the locations clustered in the Dakar area. For this, we used twitteR library in R. Using this library we scrapped tweets and extract important features from it. The features here are home location of the person (which we will obtain from CDR records), travelling destination, time of travel. In addition, we extract the interests of the person and map them to categories of interest. Again, for an example, the sample of tweets used in this work is shown in Table 2.

Table 2. Sample tweets crawled from internet.

Person identifier	Tweets	Person home location
1	Travelling to Plage de Fann to attend football match at 8 PM tomorrow	Baie de mermoz
2	Excited to attend church gathering in large Provee du club med on next Thursday	Plage de Yoff Tongor
3	Excited to meet my fiancée tomorrow morning	Baie de mermoz
4	Long time wait to meet my favorite footballer tomorrow in Plage de Fann	Plage de Yoff Tongor

Similarly, we collected all the tweets for all the users in the city. From the tweets we extract the information such as the destination and time of travel etc. Heres, it is predicted that the user travels from Baie de mermoz to Plage de Fann next day such that he/she arrives at 8 PM. Based on this, we calculate the expected time of departure and we schedule bus transport at 7.30 PM. In this way, the proposed public transport ensures maximum comfort for the user. The heat map of the travel between different points in the city is shown in Fig. 5.

From Fig. 5, the people travel between different locations in a different number. For example, it is predicted that the 235 people travel from 'Plage de Yoff Tongor' to 'Plage Privee du club Med'. Finally, these predictions are subtracted from the global model and based on these global predictions are updated. In addition to that, we observed 1000 people are planned to go for museum from the location Pikine around 10 AM. However, the global model predicted 50 people will travel from Pikine to museum at 10 AM. In this case, we update the predictions of the global model by increasing the predictions by 950 making it 1000. In this case, we schedule 50 buses from Pikine in morning 10 AM so that the people will take public transport rather than taking the cars. In this way, we can reduce pollution and encourage more people to take public transport.

Fig. 5. Heat map of the Dakar city with predictions based on social media

From the interests of the user, it is identified that 500 people are interested in football in the city. These 500 people are spread across four different regions in the city. Assume there is a football match happening in the city of Daker. Now, it is predicted that the people will visit the stadium and hence, the global model is updated with the predictions so that we can schedule effective public transport.

5 Other Applications

This paper mainly concentrates on the application of human mobility system in the field of urban planning and an effective transportation system. There are, however, a myriad of other possibilities that can be applied for the human mobility system. Crowdsourcing is one such application where the location of the person at home can be targeted to form a crowdsourcing network that provides various opportunities for employees. Mobility information can also be used to identify the low-income neighborhood and many community plans can be made that target at the betterment of the lives of the people in these regions [6]. This moreover provides various facilities which include better health facilities for the people in these regions, social development, and job training programs for the unemployed. In addition to this, a variety of applications like historic preservation, medical care, criminal justice, and traffic control systems. Location Based Services can also be provided with the help of human mobility [6]. Uber or Ola Taxi's also extend to use this type of services for building an effective transportation system related to community service development.

6 Conclusion

A good mobility system should satisfy the purpose for which it was built. The model proposed in this paper, *DHUM* is one such system, which helps in effective urban planning and transportation for metropolitan cities. Technological aspects of GIS were provided to help visualize the mobility patterns. Moreover, the greatest advantage of the proposed system as it gives real-time simulations of human mobility which can be used for a multitude of applications. With this mobility information, we schedule the public transport to accommodate the people in the city in ensuring good comfort. To further improve the scheduling, we propose to update the predictions in near real-time where we have gathered social media information of the people in the city. This ensures less pollution and maximum comfort for the people in the city. It also decreases the traffic chaos as more people travel in public transport. The approach used here will make the operates to provide better services to their customers in real-time using the exploration of dynamic data handling thus making the system more efficient and providing results at a faster rate. Innovative partnerships and business models combined with social and environmental responsibility are necessary to make the roadmap to smart cities economically, socially, and environmentally sustainable in the 21st century.

References

1. Dirks, S., Gurdgiev, C., Keeling, M.: Smarter cities for smarter growth: how cities can optimize their systems for the talent-based economy. IBM Global Business Services, Somers, NY (2010)
2. Isaacman, S., et al.: Human mobility modeling at metropolitan scales. In: Proceedings of the 10th International Conference on Mobile Systems, Applications, and Services, pp. 239–252. ACM (2012)
3. Jiang, S., Ferreira, J., Gonzalez, M.C.: Activity-based human mobility patterns inferred from mobile phone data: a case study of Singapore. IEEE Trans. Big Data 3(2), 208–219 (2017)
4. Ji, Y.: Understanding human mobility patterns through mobile phone records: a cross-cultural study. Ph.D. dissertation, Massachusetts Institute of Technology (2011)
5. Krumm, J.: Inference attacks on location tracks. In: LaMarca, A., Langheinrich, M., Truong, K.N. (eds.) Pervasive 2007. LNCS, vol. 4480, pp. 127–143. Springer, Heidelberg (2007). https://doi.org/10.1007/978-3-540-72037-9_8
6. Navidi, W., Camp, T.: Stationary distributions for the random waypoint mobility model. IEEE Trans. Mob. Comput. 1, 99–108 (2004)
7. Association of Collegiate Schools of Planning: Guide to Undergraduate and Graduate Education in Urban and Regional Planning, 17th edn. (2011)
8. Jiang, S., Ferreira, J., González, M.C.: Clustering daily patterns of human activities in the city. Data Min. Knowl. Disc. 25(3), 478–510 (2012)
9. Herder, E., Siehndel, P.: Daily and weekly patterns in human mobility. In: UMAP Workshops (2012)
10. Dwork, C.: Differential privacy. In: van Tilborg, H.C.A., Jajodia, S. (eds.) Encyclopedia of Cryptography and Security, pp. 338–340. Springer, Boston (2011). https://doi.org/10.1007/978-1-4419-5906-5_752
11. Toole, J.L., Colak, S., Sturt, B., Alexander, L.P., Evsukoff, A., González, M.C.: The path most traveled: travel demand estimation using big data resources. Transp. Res. Part C Emerg. Technol. 58, 162–177 (2015)
12. Zyba, G., Voelker, G.M., Ioannidis, S., Diot, C.: Dissemination in opportunistic mobile ad-hoc networks: the power of the crowd. In: 2011 Proceedings IEEE INFOCOM, pp. 1179–1187. IEEE (2011)
13. Dikk, D.D.: Leader Du Transport Public De Voyageurs. http://demdikk.com/. Accessed 25 Apr 2019
14. Daftary, K., Kapadia, R., Prajapati, D., Shirole, M.: Routed: a dynamic bus scheduling system. In: IEEE Symposium Series on Computational Intelligence (SSCI), Bangalore, India, pp. 74–82 (2018)
15. Newell, G.F.: Dispatching policies for a transportation route. Transp. Sci. 5(1), 91–105 (1971)
16. Hall, R., Dessouky, M., Quan, L.: Optimal holding times at transfer stations. Comput. Ind. Eng. 40(4), 379–397 (2001)

Effects of Consumption Value on Online Repurchase Intention: Mediation Effect of Green Information Visibility

Jia-Hong Lim[1], Jung-Chen Chen[2], and Chien Hsing Wu[2(✉)]

[1] International Business of Administration, National University of Kaohsiung,
Kaohsiung City, Taiwan
reaper0.0@hotmail.com

[2] Department of Information Management, National University of Kaohsiung,
Kaohsiung City, Taiwan
ml073309@mail.nuk.edu.tw, chwu@nuk.edu.tw

Abstract. The purpose of the paper is to propose and examine empirically a research model that describes the online repurchase intention. The proposed model is based on the concepts of expectation-confirmation theory and consumption value theory that includes internal, external, and functional values as the determinants of online consumer's buying satisfaction towards online repurchasing intention. The green information visibility as a mediator to influence the effect of consumption values on consumer purchasing satisfaction is also examined. Based on the analysis of 308 valid samples, the main research findings are as follows. (1) Among those three values, internal and functional value highly influence their satisfaction. (2) Functional value is the only variable that is affected by the mediator while internal and external are not affected. (3) For sub-factors, quality and usability strongly support consumer's satisfaction while price and attitude partially support their satisfaction. (4) Attitude, price and quality value has mediating effect except the others. Discussion and implications will also be delineated.

Keywords: Consumption value · Green information visibility · Consumer electronic products · Online repurchase intention

1 Research Background

Past purchase or taking order for goods are generally by making phone call or walk into a store. However, the innovation of internet transfers the method of making a trade into online visual store. This innovation gives a lot of benefits, no matter at consumer or supplier side. Online transaction not only increases the number of trading, but also provides a convenient platform to make a trade between host country and client countries [17]. The online platform is friendly to use and also provides information and picture of the product that let us have multiple choices on the product types and suppliers. However, online marketers have paid increasing attention to the factors that may be used to describe repurchase satisfaction towards repurchase intention. Hence, a research question may be raised that what are the main determinants that are associated with the online repurchase or not repurchase behavior.

© Springer Nature Singapore Pte Ltd. 2019
J. C.-W. Lin et al. (Eds.): MISNC 2019, CCIS 1131, pp. 43–55, 2019.
https://doi.org/10.1007/978-981-15-1758-7_4

Literature indicated that repurpose behaviors involve various factors ranging from economic, individual, to psychological appeals. For example, Koo and Ju [1] tested the effect of atmospheric cues of online stores on the intervening affective emotional states of consumers, which has a subsequent impact on behavioral intention. Tsai el at. [2] indicate that whether a more prominent display of privacy information will cause consumers to incorporate privacy considerations into their online purchasing decisions; William et al. [3] present a comprehensive review of recent empirical studies dealing with online consumer behavior and decision-making processes. These studies discuss about the effect of external value, internal value and privacy information on online purchasing behavior. However, the effect of green information visibility on the links of consumption value and satisfaction has not been discussed.

Furthermore, searching and examining the information which explains the feature, functional and manipulation of particular product is one of the key factors that influence consumer purchasing behavior [14, 18]. In this study, the green information visibility is used as a mediator which affects the consumers to repurchase electronic products. In fact, when choosing between two similar products, according to a survey from DoubleClick Performics, 83 percent of consumers are extremely or very likely to choose the environmentally friendly option, that gauges online consumer behavior and attitudes regarding green marketing. This implies that consumer will be willing to purchase product with green information.

Generally, consumers would like to be a smart consumer rather than a compulsive or susceptible consumer. To keep thinking the impact or benefit of the product is what we usually do before making a decision. Therefore, consumers should consider many variables, such as culture, social, usability, etc. From the perspective of psychology of consumers, exploring what they care mostly will be likely benefit to the marketing strategies because focusing on what they can be touched will greatly contribute to satisfaction level towards purchasing products again and again. Moreover, consumers are paying increasing attention to the concepts of environment protection. In consequence, products along with green information are being highlighted, in particular the online goods. The research thesis is motivated by two issues and reaches two main objectives, (1) to propose and empirically examine a research model that incorporates both consumption value and expectation and confirmation theory to describe repurchase satisfaction towards repurchase intention in the context of online consumer electronic products, (2) to examine the mediation effect of green information visibility on the relationship between consumption value and repurchase satisfaction.

2   Related Concepts and Hypothesis Development

2.1   Expectation-Confirmation

This paper examines human's psychology and functional effect influencing one's intention to continue purchasing and using the electronic product. Expectation-confirmation theory (ECT) is adapted from literature and integrated with theoretical and empirical findings from prior consumption value research to theorize a model of consumer repurchase intention of electronic products.

The ECT is from prior IS usage research which is widely used in the consumer behavior literature to study consumer satisfaction, post-purchase behavior (e.g., repurchase, complaining), and service marketing in general by Bhattacherjee [4]. The predictive ability of this theory has been demonstrated over a wide range of product repurchase and service continuance contexts, including automobile repurchase, camcorder repurchase, institutional repurchase of photographic products, restaurant service, and business professional services. In general, the ECT explains that consumers' intention to repurchase a product or continue service use is determined primarily by their satisfaction with prior use of that product or service by Bhattacherjee [4].

Generally, when expectations are low and experiences are high, individuals may adjust their satisfaction evaluation downward towards their expectations, thus enacting a dissonance reduction strategy. Moreover, lower satisfaction levels will be their ultimate evaluation, which is influenced by their preconceived expectations. Similarly, when expectations are high and experiences are low, individuals may adjust their satisfaction evaluations upward towards their expectations. In this case, the positive effect of experiences is actually enhanced by positive expectations, such that when individuals have positive expectations and positive experiences their satisfaction is likely high by Susan el at. [5]. The expectation is usually based on others' experience or information obtained through product reviews or word of mouth.

2.2 Consumption Value Theory

The consumption value theory refers to subjective beliefs and desirable attitude that will become personal values. People generally achieve personal values through actions or activities, such as social interaction, economic exchange, possession, and consumption by Sheth et al. [6]. According to means-ends chains models of consumer product knowledge by Peter & Olson [16], people may have ideas and preferences about various actions that can help them achieve personal values. Therefore, the creation of personal values from a product depends on consumption values which are instrumental in nature. For example, owning an elegant house and acquiring a prestigious car are for some people to the desirable ways of achieving self-fulfillment. Attending football games (especially those of favorite teams) and taking a vacation trip are favorable activities which lead to personal fun and enjoyment. Furthermore, individuals may develop several personal values by which they direct or evaluate consumption activities [15]. Therefore, the consumption values of these types of activities (or possessions) are sophisticated and do not simply satisfy one single personal value by Sheth et al. [6]. These values may include epistemic value, social value, price value, etc. In order to clarify these values, they are distributed into three categories: internal value, external value, and functional values.

2.2.1 Internal Value
When confronted with a purchase decision, we may engage in internal search by scanning our own memory bank to assemble information about different product alternatives. Usually, though, even the most market-aware of us needs to supplement this knowledge with external search, where information is obtained from advertisements, friends, or just plain people-watching by Solomon et al. [7]. The inspiration of these information may

directly make an impression in our own memory bank that will shape some perspective to the particular product such as epistemic value and attitudes value.

2.2.2 External Value

During the consumer decision-making process, internal memory is combined with external memory – which includes all the product details on packages in shopping lists, and through other marketing stimuli – to permit brand alternatives to be identified and evaluated by Solomon et al. [7]. This value is created by consumer's external memory to value a product. Thus, there are many external factor that may influence consumer memory, such as fashion, culture, environment, etc. To specify the factors, the research chooses social value, education value, and culture value as external value to evaluate consumer purchase behavior.

2.2.3 Functional Value

Other than the previous value, there is another value which is necessary to be included, functional value. The functional value of an alternative is defined as the perceived utility acquired from an alternative's capacity for functional, utilitarian or physical performance. For example, the decision to purchase a particular automobile might be based on fuel economy and maintenance record by Sheth et al. [6]. Thus, to find out the purchasing behavior of consumer, functional value will involve four attribute: price, quality, attribute, and usability in this research.

Generally, the price as the first factor on goods that we are willing to purchase or trade in, is a part of the marketing mix, used to stimulate the consumer and become a communicator, to negotiate and a competitive weapon. Moreover, when consumers try their best for a lower price, those consumers are defined as price-sensitive. Planned future price increases should be communicated to consumers as well as planned price cuts, since it gains from the decreased uncertainty more than offsets the loss from the delay in purchase behavior.

Another factor is needed to consider, that is quality factor. In Zeithmal's [8] concept, quality was seen as the superiority of a certain product or service in a broad sense. Perceived quality was defined as consumers' judgment about products', excellence or superiority. In particular, the online market for experienced products provides a substantial and largely untapped revenue source, but represents a challenge for online sellers, as the technology-mediated environment makes it more difficult to convey the experiential attributes associated with such products (e.g., taste, sound, fit). While some researchers are investigating how a virtual product experience (VPE) provided through a Web interface can better convey visual product attributes the e-commerce channel remains limited in conveying experiential attributes as compared to a physical store.

Usability includes three main concepts: effectiveness, efficiency, and user consummation. In its broadest sense, usability is the expression of whether the product is good enough to meet all needs of the user. User evaluation of the usability measures affects purchasing of the product. Han et al. [9] stated that image/impression and performance dimensions should be considered together to evaluate usability of electronic devices. To be able to show the effects of interface factors, performance and image/impression dimensions are detailed by Kumru & Ergun [10]. In order to measure the level of consummation of the user, as well as effectiveness and efficiency, usability testing is needed.

2.3   Green Information Visibility

The problem of information visibility on the Internet is one that is primarily attributed to anonymity, lack of face-to-face interactions, unfamiliarity between buyers and sellers and the lack of repeated interaction. One way to mitigate online risk is via information visibility, which has shown to help to lower user risk belief. Consumers generally felt they received insufficient information about products and services, and hence they preferred more information and details, not only on pricing and services, and on production processes and health impact as well. The provision of appropriate information by organizations is seen as building trust with consumers and facilitating perceptions of reliability by Koo et al. [11].

From an academic perspective, the area of "Green Information Technology" has been rigorously studied [11–13]; however, the trend toward using technologies, innovation, and techniques requires further exploration. This fairly new area of inquiry is referred to as "Green IT," a term applied to the most important strategic technology to better support sustainable practices; the term also closely relates to information systems and business processes. Watson et al. [12] criticized the fact that the IS academics primarily have focused on productivity improvements and economic efficiency rather than supporting environmental practices. This focus has resulted in waste, unused resources, energy inefficiency, noise, friction, and emissions. There is a lack of information regarding how to bring about economic benefits while encouraging social and ecological responsibility for the future generations by Koo el at. [11].

While environmental sustainability has received attention amongst consumers, Green IT has just begun to display its unique practical contributions, both negative and positive, to the environment and economy. Green IT & IS will necessarily include both organizational and individual aspects of environmental technology (people, processes, software, and information technologies) by Sarkis et al. [13]. Similarly, the research argument for this issue is whether visibility of green information mediates the effect of consumption value on satisfaction towards repurchase intention.

3   Method

3.1   Research Model

This research paper focuses on examination of the consumer online repurchase intention of consumer electronic product. Based on the review of literature and the arguments of the current research focus, the research model is presented in Fig. 1, which contains three parts. The first part has three independent variables including internal value, external value and functional value. The internal value contains sub-variables of epistemic and attitudes. The external value has social, educational, and culture while functional value includes price, quality, attribute, and usability. The second part is the independent variables that have satisfaction and online repurchase intention. The satisfaction is also regarded as the independent variable of online repurchase intention. The final part is the mediator, green information visibility, which is included in the model to alternate the effect of independent variables on dependent ones. Hypotheses are defined as follows.

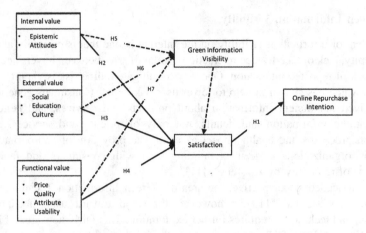

Fig. 1. Research Model

3.2 Sampling Plan

The research subject who have buying experience or willing to buy consumer electronic product in the future that will be focused on. Next, the questionnaire will ask the subjects about if the green information visibility of the product influence their satisfaction when they purchased it. And also, in order to investigate about the consumer's online repurchase intention, subjects will be asked in the questionnaire whether they satisfy the e-commerce platform that leads them to purchase from virtual store. Then, the questionnaire is sent to the subjects through the internet and handout. The sample target is college student, friends, family, relative or colleague and will expect to have the responders more than 300 people. The software for setting up the questionnaire is used via Google drive and internet social media such as Facebook and others for easily sharing. In terms of gender and age, there should be not too much gap that this result is able to reflect the idea of the general consumers.

3.3 Measure

The survey questionnaire contains two main parts. The first is questionnaire to inquiry the subjects' basic information, such as age, gender, salary or willing to pay. Another part is about the subject's personal opinion about the question items. For instance, the question will ask subject how variable impacts them after they consume or try the electric product, do they feel fulfillment and desire what the value was provided by the consumer electronic product. Due to the questionnaire was designed to ask what the feeling of the subject from consumer electronic product consumption; the answer may request to fill on 1 to 7, Likert 7-scale instrument. The 1 represents the subject does not find out or a little sense perception towards the variable by the consumer electronic product, on while 7 defines as the subject having a highest sense perception towards the variable from the consumer electronic product. Using number for answering the questionnaire can let the report easily understand and discriminate the subjects' mind set; and at the same time, the statistics are possible to derive the research findings.

4 Results

4.1 Descriptive Statistics

The total valid questionnaires is 308, having 128 for male and 180 for female. Disposable income varies from less than 5000 to more than 20000TWD. Consumption experience ranges from less than 3 years to more than 6 years. The subjects are also asked to choose a brand and three accessories they preferred. Results are presented in Fig. 2 and 3. This accounts the use of CD/DVD player is getting least and least contrast to smartphone which has over 200 votes from the subjects.

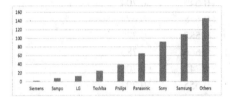

Fig. 2. Brand preference for the respondents　　**Fig. 3.** Accessories preference for the respondents

4.2 Reliability and Factor Analysis

To ensure that the reliability are consistent and stable for the observational data, the Cronbach's Alpha coefficient of 13 components for 9 independent variables factors, 2 dependent variables and 1 mediator were analyzed.

In the distribution of components, independents of epistemic value and attitudes value are under internal value; social value, educational value and culture value belong to external value; price, quality, attribute and usability are classified as functional value. Then, Cronbach's Alpha of each factor, item to total and Cronbach's Alpha if item deleted are showed in Table 1. It is noticed that FV7, 8, and 9 were deleted due to low reliability, which were not considered in the factor analysis. The factor analysis results for independent variables and dependent are presented in Table 2 and 3, respectively.

Table 1. Cronbach's Alpha for independent variables

Ind. variables	Items	Cronbach's Alpha	Item to total	Cronbach's Alpha if item deleted
Epistemic value	IV1	0.698	0.459	0.672
	IV2		0.525	0.594
	IV3		0.562	0.545
Attitudes value	IV4	0.620	0.259	0.795
	IV5		0.639	0.249
	IV6		0.451	0.491
Social value	EV1	0.873	0.773	0.806
	EV2		0.751	0.825
	EV3		0.744	0.831
Educational value	EV4	0.815	0.648	0.766
	EV5		0.657	0.755
	EV6		0.697	0.717
Culture value	EV7	0.792	0.649	0.704
	EV8		0.678	0.674
	EV9		0.587	0.779
Price	FV1	0.829	0.666	0.783
	FV2		0.742	0.706
	FV3		0.655	0.794
Quality	FV4	0.815	0.547	0.884
	FV5		0.758	0.656
	FV6		0.720	0.696
Attribute	FV7	0.543	0.313	0.511
	FV8		0.333	0.474
	FV9		0.421	0.333
Usability	FV10	0.773	0.606	0.698
	FV11		0.701	0.589
	FV12		0.526	0.786
Satisfaction	SA1	0.901	0.846	0.847
	SA2		0.865	0.841
	SA3		0.685	0.906
	SA4		0.728	0.891
Online repurchase intention	ORI1	0.947	0.844	0.939
	ORI2		0.874	0.930
	ORI3		0.902	0.921
	ORI4		0.869	0.931
Green information visibility	GIV1	0.861	0.695	0.842
	GIV2		0.751	0.791
	GIV3		0.764	0.778

Table 2. Factor analysis results for independent variables

Items	Factor loadings						
	1	2	3	4	5	6	7
FV6	**0.732**	0.015	0.209	0.269	0.328	0.204	0.025
FV5	**0.709**	0.095	0.113	0.336	0.313	0.178	0.036
FV12	**0.684**	0.164	0.001	−0.037	0.151	0.161	0.343
FV11	**0.672**	0.010	0.218	0.116	0.152	0.130	0.437
FV10	**0.632**	−0.115	0.350	0.333	0.218	0.070	0.117
EV1	0.042	**0.895**	−0.007	−0.048	0.136	0.022	0.012
EV3	0.038	**0.864**	−0.074	0.009	0.106	0.113	−0.044
EV2	0.025	**0.862**	0.063	0.136	0.066	0.201	0.043
EV9	−0.022	−0.034	**0.800**	0.207	0.126	0.004	0.227
EV7	0.287	−0.029	**0.731**	0.115	0.068	0.383	0.020
EV8	0.287	0.009	**0.727**	0.100	0.143	0.192	0.175
IV2	0.092	−0.012	0.071	**0.745**	0.186	0.282	0.053
IV3	0.235	−.052	0.138	**0.711**	0.101	0.167	0.276
IV1	0.234	0.203	0.314	**0.580**	0.022	0.080	0.120
FV2	0.313	0.159	0.091	0.141	**0.808**	0.184	−0.027
FV3	0.237	0.129	0.074	0.026	**0.756**	0.154	0.332
FV1	0.284	0.207	0.342	0.282	**0.646**	0.028	0.096
EV6	0.177	0.167	0.342	0.170	0.055	**0.764**	0.086
EV4	0.164	0.164	0.035	0.169	0.260	**0.763**	0.169
EV5	0.193	0.181	0.176	0.424	0.078	**0.633**	0.160
IV6	0.204	−0.022	0.251	0.208	0.152	0.212	**0.762**
IV5	0.396	−0.004	0.257	0.324	0.127	0.121	**0.659**

1. Extraction Method: Principal component analysis
2. Rotation Method: Varimax with Kaiser Normalization
3. IV: Internal value; EV: External value; FV: Functional value

Table 3. Factor analysis for the dependent variables

	Component	
	1	2
ORI3	**0.938**	0.136
ORI2	**0.922**	0.117
ORI4	**0.920**	0.127
ORI1	**0.899**	0.144
SA2	0.117	**0.926**
SA1	0.120	**0.915**
SA4	0.076	**0.846**
SA3	0.187	**0.791**

1. Extraction Method: Principal component analysis
2. Rotation Method: Varimax with Kaiser Normalization
3. SA: Satisfaction; ORI: Online repurchase intention

4.3 Regression Analysis

The multiple regression model were used to test the hypotheses. Results are presented in Table 4 for H1, Table 5 for H2, H3, and H4, Table 6 for significant effect of green information visibility on satisfaction, Table 7 for significant effect of internal, external and functional value on green information visibility, and Table 8 for the mediation test results (H5, H6 and H7). Moreover, the test results for the H2a, H2b, H3a, H3b, H3c, H4a, H4b and H4d are revealed in Table 9.

Table 4. Test Results for H1

Dep. variable	Online repurchase intention			
Ind. variable	R^2	Beta	t	p-value
Satisfaction	0.078	0.280	5.104***	<0.0001

Table 5. Test results for H2, H3 and H4

Dependent variable	Satisfaction			
Ind. variable	R^2	Beta	t	p-value
Internal value	0.537	0.218	3.897***	<0.0001
External value		0.067	1.305	0.193
Functional value		0.609	10.791***	<0.0001

Table 6. Test results for mediator

Dep. variable	Satisfaction			
Ind. variable	R^2	Beta	t	p-value
Green information visibility	0.109	0.33	6.114***	<0.000

Table 7. Tests results for internal, external and functional value towards green information visibility

Dependent variable	Green information visibility			
Ind. variable	R^2	Beta	t	p-value
Internal value	0.163	0.117	1.562	0.119
External value		0.225	3.282***	<0.001
Functional value		0.309	4.066***	<0.0001

Table 8. Test results for H5, H6 and H7

Dep. variable	Satisfaction					
	Model 1			Model 2		
Ind. variable	Beta	t	p-value	Beta	t	p-value
IV	0.218	3.897***	<0.001	0.229	4.117***	<0.000
EV	0.067	1.305	0.193	0.089	1.723*	0.086
FV	0.609	10.791***	<0.0001	0.579	10.056***	<0.0001
GIV				0.099	2.330**	0.02
R^2	0.537			0.546		

IV: Internal value; EV: External value; FV: Functional value; GIV: Green information visibility

Table 9. Test results for H2a, H2b, H3a, H3b, H3c, H4a, H4b and H4d

Dep. variable	Satisfaction			
Ind. variable	R^2	Beta	t	p-value
Epistemic	0.56	0.054	1.003	0.317
Attitudes		0.121	2.129**	0.034
Social		0.067	1.560	0.120
Educational		0.009	0.168	0.866
Cultural		0.040	0.805	0.421
Price		0.113	2.064**	0.04
Quality		0.229	3.654***	<0.001
Usability		0.336	5.198***	<0.001

5   Discussion and Conclusion

This research's purpose is to find out what factors affect consumer repurchasing online electronic product by using the consumption value as the model to conduct a thorough analysis. The green information visibility was used to test the mediating impact on consumption behaviour. Some important findings are addressed after analysis of the samples and obtaining the results as follows:

First, consumers were hypothesized that they are impacted by internal, external and functional value when they were buying consumer electronic products. Among these three values, internal and functional value highly influence their satisfaction, but external value does not impact their satisfaction. Second, there are total seven sub-factors that were tested toward the satisfaction. Quality and usability are strongly supported to be influential to satisfaction while price and attitude partially support their satisfaction. The remaining variables, epistemic, social, education and cultural are not supported in this model. Third, External value is only variable that is not affected by the mediator while internal and functional are significantly affected. Finally, attitude, price and quality value has mediating effect instead of the others.

With the advance of science and technology, technology firm will keep producing innovative, and replace old consumer electronic products. This study can be a reference for future study on related industry's research. In order to better understand consumer purchasing behaviour, combining the consumption value theory with consumer's satisfaction and repurchase intention to get a stronger perspective from post purchase experience and perceived satisfaction of consumer that tend to online repurchase intention. From the standpoint of consumer electronic industry, the results illustrate that consumer's satisfaction highly and significantly relate to internal and functional value. To attract consumer's attention and increase the satisfaction of consumer, firms are suggested to focus more on internal value and functional value. For example, retailer can exposes more information about the innovation or accessible of new product through the advertisement or social media.

Similar topic of future studies could be given some directions from this research findings and suggestions. Consumers are more willing to purchase product from internet. So, there are others products such as, clothes, daily essentials and etc. which are interesting to know how consumer behaviour will be affected by them. Different kinds of consumer electronic products will have different kinds of features. In order to find out the differentiation, they can separately be discussed in the future study. This can be more specified to address each of them of value creation for consumers to increase satisfaction. Moreover, making a more precise distinction on the demographic statistics or target market to analysis the consumer behaviour that will get a better understanding between the variables and certain product.

References

1. Bhattacherjee, A.: Understanding information systems continuance: an expectation-confirmation model. MIS Q. **25**(3), 351–370 (2010)
2. Koo, C., Chung, N., Nam, K.K.: Assessing the impact of intrinsic and extrinsic motivators on smart green IT device use: reference group perspectives. Int. J. Inf. Manag. **35**, 64–79 (2015)
3. Koo, D.M., Ju, S.-H.: The interactional effects of atmospherics and perceptual curiosity on emotions and online shopping intention. Comput. Hum. Behav. **26**, 377–388 (2010)
4. Slade, E.L., Williams, M.D., Dwivedi, Y.K.: Mobile payment adoption: classification and review of the extant literature. Mark. Rev. **13**(2), 167–190 (2013)
5. Marriott, H.R., Williams, M.D., Dwivedi, Y.K.: What do we know about consumer m-shopping behavior? Int. J. Retail. Distrib. Manag. **45**(6), 568–586 (2017)
6. Suraiya, I.I., Zabil, N.F.M.: Impact of consumer awareness and knowledge to consumer effective behavior. Asian Soc. Sci. **8**(13), 108–114 (2012)
7. Sarkis, J., Koo, C., Watson, R.: Green information systems & technologies—this generation and beyond: Introduction to the special issue. Inf. Syst. Front. **15**(5), 695–704 (2013)
8. Peter, J.P., Olson, J.C.: Consumer Behavior and Marketing Strategy, 2nd edn. Irwin, Homewood (1990)
9. Sheth, J.N., Newman, B.I., Gross, B.L.: Why we buy what we buy: a theory of consumption value. J. Bus. Res. **22**, 159–170 (1991)
10. Tsai, J.Y., Egelman, S., Cranor, L., Acquisti, A.: The effect of online privacy information on purchasing behavior: an experimental study. Inf. Syst. Res. **22**(2), 254–268 (2011)

11. Atalay, K.D., Eraslan, E.: Multi-criteria usability evaluation of electronic devices in a fuzzy environment. Hum. Factors Ergon. Manuf. Serv. Ind. **24**(3), 336–347 (2014)

12. Shareef, M.A., Dwivedi, Y.K., Kumar, V., Kumar, U.: Content design of advertisement for consumer exposure: mobile marketing through short messaging service. Int. J. Inf. Manag. **37**(4), 257–268 (2017)

13. Solomon, M., Bamossy, G., Askegaard, S., Hogg, M.K.: Consumer Behavior: European Perspective, 3rd edn. Financial Times and Prentice Hall, Upper Saddle River (2006)

14. Watson, R.T., Boudreau, M.-C., Chen, A.J.: Information systems and environmentally sustainable development: energy informatics and new directions for the IS community. MIS Q. **34**(1), 23–38 (2010)

15. Han, S.H., Yun, M.H., Kwahk, J., Hong, S.W.: Usability of consumer electronic products. Int. J. Ind. Ergon. **28**, 143–151 (2001)

16. Brown, S.A., Venkatesh, V., Kuruzovich, J., Massey, A.P.: Expectation confirmation: an examination of three competing models. Organ. Behav. Hum. Decis. Process. **105**, 52–66 (2002)

17. Zeithmal, V.A.: Consumer perceptions of price, quality, and value: a means-end model and synthesis of evidence. J. Mark. **52**, 2–22 (1988)

18. Darley, W.K., Blankson, C., Luethge, D.J.: Toward an integrated framework for online consumer behavior and decision making process: a review. Psychol. Mark. **27**(2), 94–116 (2010)

An Effective BI-encoded Schema for Mention Extraction

Jerry Chun-Wei Lin[1], Jimmy Ming-Tai Wu[2(✉)], Yinan Shao[3], Matin Pirouz[4], and Binbin Zhang[5,6]

[1] Department of Computer Science, Electrical Engineering and Mathematical Sciences, Western Norway University of Applied Sciences, 5063 Bergen, Norway
`jerrylin@ieee.org`
[2] College of Computer Science and Enginnering, Shandong University of Science and Technology, Qingdao 266, China
`wmt@wmt35.idv.tw`
[3] School of Computer Science and Technology, Harbin Institute of Technology (Shenzhen), Shenzhen 518055, China
`shaoyn0817@163.com`
[4] Department of Computer Science, California State University, Fresno, CA 93740, USA
`mpirouz@ieee.org`
[5] Department of Biochemistry and Molecular Biology, Shenzhen University Health Science Center, Shenzhen 518055, China
`zhangbb@szu.edu.cn`
[6] Center for Anti-aging and Regenerative Medicine, Shenzhen University Health Science Center, Shenzhen 518055, China

Abstract. We present a neural-encoded mention-hypergraph (named as NEMH in this paper) model for mention-extraction and classification in this paper. Through extraction of textual mention entities, a model is proposed that applies a hypergraph-encoding schema to neural networks. Comparing the results of the proposed model with the previous approaches, the proposed model can thus identify unlimited-length nested mention entities, which is a major milestone in the field. Several experiments are conducted on many datasets used in the baseline approaches, and the obtained results indicated that the designed model has high effectiveness compared to the existing models.

Keywords: Mention hypegraph · Sequence labeling · Bi-LSM · Neural network

1 Introduction

Mention extraction [8] can thus be referred to the process of identifying mention-entities and tagging them with labels. This procedure is similar to those classic tasks of sequence labeling, as shallow parsing or name entity-recognition. Sequence labeling proceeds with assigning semantic labels to texts of varying

© Springer Nature Singapore Pte Ltd. 2019
J. C.-W. Lin et al. (Eds.): MISNC 2019, CCIS 1131, pp. 56–67, 2019.
https://doi.org/10.1007/978-981-15-1758-7_5

lengths. Therefore, the segmented text is considered a sub-sequence of a given label. For an text input of length n, the theoretical text span is $[1, n]$.

Mention-extraction is a popular research topic among researchers because of its importance in various downstream tasks such as co-reference resolution [13], entity linking [22], and relation extraction [12,20]. Although these models can be applied to the task of mention-extraction, overlapping mention entities cannot be effectively identified.

In a study conducted by Finkel and Manning [9], a discriminative parser was presented as tree-based structure which identifies nested labels. The proposed framework achieves promising results but at the cost of a large time complexity. Lu and Roth [17] utilized hypergraph, and managed to improve the time complexity to $O(n)$. Later, Muis and Lu [21] performed mention extraction by applying multi-graph representation to mention separators. Inspired by previous works, this work studies a hypergraph model considering the built mention using neural networks with varying encoding schemes to increase the performance and usability. The proposed encoded hypergraph model uses neural networks to facilitate feature score extraction for the given edges/hyperedges. The significant contributions of this paper are described below.

- A neural-encoded-based mention hypergraph (also named as NEMH in this paper) model is first presented to conceive the nested-structure mention entities.
- Moreover, the hypergraph-based model uses BI (**B**egin, **I**nside) encoding schema, in order to obtain more boundary features than the existing works.
- Through experimentation, we found that the designed model outperforms existing approaches when tested on benchmark sequence prediction datasets with a higher effectiveness than the existing works.

2 Related Work

Sequence prediction has been performed using the hidden Markov model (HMM) [2–6], conditional random fields (CRF) [18], max-entropy model (MEM) [1], and semi-Markov random fields (semi-CRF) [23]. Leonard et al. [2–6] modeled a HMM using a dynamic Bayesian network. Moreover, Fine et al. [11] developed a hierarchical HMM to generalize the HMM as a recursive hierarchy.

Berger et al. [1] developed a MEM for handling the NLP tasks. They presented a maximum-likelihood model, which can be used to automatically construct maximum entropy models. This efficient approach can be used to estimate the parameters, and can be utilized in some real-applications, specifying for the natural language processing tasks. A follow-up work presented by McCallum et al. [19] studied the task of sequence prediction by proposing a Markov-based MEM that enables a graphical modeling of the problem which benefits from characteristics of both HMM and MEM. Markov-based MEM allows the representation of observations as random overlapping features (namely words, capital letters, formatting, parts of speech). Those approaches can be used to model the conditional probability of state sequences based on observation sequences by the maximum entropy framework. The objective is to find a set of exponential models. Those models can thus be utilized to describe the probability of a state by an observation, as well as the elder state.

Lafferty et al. [18] in 2001 developed conditional random field (CRF) models that are statistically modeled and utilized the problem for predicting the sequences. The advantages of this approach over HMM and stochastic grammars include relaxation of strong independence assumptions. Challenges such as bias toward states with few successor states are addressed by CRF, which differentiates CRF from maximum entropy Markov models and other discriminative Markov-based models targeting to the directed graphical models. Finkel et al. [10]'s feature-rich discriminative parser model benefits from the CRF characteristics and fit the full Wall Street Journal (WSJ) dataset. Their proposed model uses stochastic optimization techniques, as well as parallelization and chart pre-filtering to outperform the existing methods.

Sarawagi and Cohen [23] proposes a semi-Markov CRF (semi-CRF). Through experimentation on named entity-recognition problems, they showed that semi-CRF generally outperforms conventional CRFs.

More recently, deep learning methods are proving promising in sequence labeling. Huang et al. [15] tackled sequence prediction using long short-term memory (LSTM). In a recent study, Dyer et al. [7] modeled a stack LSTM, which is developed as a transition-based dependency parser. Their framework identifies parser states directly. Furthermore, Lample et al. [16] introduced a pair of models for sequence labeling. Their proposed models use bi-directional LSTMs (generally named as the Bi-LSTMs) alongside conditional random fields (CRFs), and transition-based to construct and tag the segments, respectively. Their models are mostly based on bi-information referring to words, i.e., representation for the character-based word that is learned from the supervised corpus. The unsupervised word representation can be learned from the unannotated corpora. From the experimental results, those models can obtain the better performance than the elder works in the task of NER without resorting any of the language-specific knowledge or resources. Zhuo et al. [24] developed gated recursive semi-CRFs, which uses a gated recursive CNN to model and learn segment-level features.

The existing research on the mention-extraction provides much insight into the state of the range of possibilities in this field. Finkel and Manning [9] based their work on a discriminative constituency parser, which builds on the tree structure to identify nested-name entities. From this developed model, it has obtained $O(n^3)$ in terms of time complexity, as a function of the number of words in the input sentence. Muis and Lu [21] then utilized mention separators and multigraph representation for extracting mention entities. This model is capable of extracting not only nested structures but also overlapping structure-mention entities. Their proposed framework achieves a decent time complexity in the studied experimentation.

3 Preliminary and Problem Statement

The preliminaries used in this paper are introduced below.

3.1 Recurrent Neural Networks (RNN)

Recurrent neural networks (RNN) are a popular variation of neural networks, that perform well for sequential data. LSTM-based neural networks [14] are the specific category of RNN, which outperform the other architectures when it comes to long input. The underlying architecture of a given LSTM unit is comprised of three multiple gates that can be used to respectively control the amounts of detailed discarded and passed onto the next epoch. The generic LSTM cell is then showed in Fig. 1.

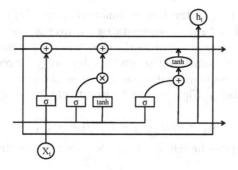

Fig. 1. The generic LSTM cell.

We utilize Bi-LSTM [5] as shown in Fig. 2 to the problem of sequence labeling prediction. The nodes marked as the round at the bottom is used to represent the vectors for the input, the nodes marked as the square at the top is used to represent the vectors for the output, and the nodes marked as the rectangular at the middle is used to represent the LSTM units (from Fig. 1). The Bi-LSTM moves forward and backward sequences to two separated LSTMs and the output is formed by concatenating the respective outputs of the LSTM neural networks at each epoch.

3.2 Hypergraph Model

Hypergraph models generalize a class graph model. For the classic CRF-based model, each edge combines a single node only. In contrast, the hypergraph model models with one (hyper-)edge having connections to a random number of nodes. Here, a hypergraph-based models are the models with conditional probability of a given output sequence s for a given input sequence x:

$$p(s|x) = \frac{1}{Z(x)} exp\{W \cdot G(x, s)\}, \tag{1}$$

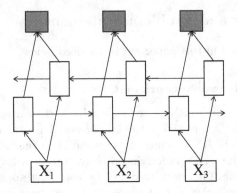

Fig. 2. The standard Bi-LSTM architecture.

in which $G(x, s)$ represents the feature function. The $Z(x)$ stands for the normalization aspect of the set of segmentations s over x, and W represents the weight vector. A hypergraph identifies the closest label sequence through α_j representing the best label sequence that ends with input j. Moreover, the (m, n, y) represents a labeling sequence with position m as the start, and n-th position as the end with a label y. Then α_j can be iteratively found by:

$$\alpha_j = max_y \psi(j - 1, j, y) + \alpha_{j-1}, \tag{2}$$

in which $\psi(j - 1, j, y)$ can be referred to the feature value to the edge $(j - 1, j, y)$.

3.3 Neural Hypergraph Model

A neural hypergraph network is based on the capable of identifying global context characteristics with the help of an LSTM layer. This can be expanded to other features as sought by a given user such as transition features or n-gram features through a hypergraph layer. The proposed model considers numerous features as detailed in Subsect. 4.3).

As an example to further explain the proposed model, we use the tag-transition feature. Assume a matrix containing tag-transitions is denoted as $[A]$, in which each $[A]_{i,j}$ can be used to represent the transition score from i to j at each epoch. Note that the transition matrix does not depend on the location. $f_\theta([x]_i^T)$ represents the feature score that created by the θ (can be referred to a neural network).

For an input sentence $[x]_i^T$, each element $[f_\theta]_{i,t}$ of the matrix can be considered as the transition score of i for a word t. For an input sentence, i.e., $[x]_i^T$, a given output label path $[i]_i^T$ is considered to be the sum of transition and neural network scores such as:

$$s([x]_i^T, [i]_i^T, \theta) = \Sigma_{t=1}^{T}([A]_{[i]_{t-1},[i]_t} + [f_\theta]_{[i]_t,t}) \tag{3}$$

Problem Statement: For the input of length represented as k $x = (x_1, \ldots, x_k)$, assume $x_{a:b}$ is the sub-sequence $(x_a, \ldots, x_b)$, in which $a \leq b \leq k$. Thus, we can define that a mention entity is considered as the triad (u, v, y) that is the sub-sequence $x_{u:v}$ and associated with mention entity label y. For a given input sequence, i.e., x, the mention-extraction problem can be modeled as the input sequence x that all the mention entities are identified, which can be either overlapping or nested.

4   Proposed Model of Mention Hypergraph

The mention hypergraph model [17] utilized both the nodes and directed hyper-edges to encode mentions regarding different natures. Figure 3 presents a partial mention hypergraph, where the label paths are given as the input sentences. There are five categories of nodes:

- A_k which presents every mention starting at or after k
- E_k which presents every mention left-bound at k
- T_k^j which presents every mention left-bound at k for type j
- I_k^j which presents every mention at k for type j
- X which presents the end of the mention.

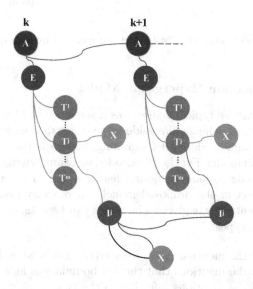

Fig. 3. The part of the mention hypergraph model (or called partial mention hypergraph model).

From the partial mention hypergraph model of Fig. 3, each A_k node is coupled with the A_{k+1} and E_k nodes. This is done to find the reason behind having the set of mentions that is started at or after k as the union of the set of mentions, started at or after $(k+1)$, and those started at k. For each E_k node is coupled with $(T_k^1, T_k^2, \ldots, T_k^m)$ via a hyperedge, in which m is considered as the number of mention types. For a hyperedge, it indicates that mentions at k is a type m. Each T_k^j node has an edge pair, which is linked to: (1) an I_k^j node, with type T^j starting at k; or (2) an X node, which does not match a type j starting at k. For every node I_k^j, there are three edges toward: (1) an I_{k+1}^j node, which represents that mention j spans over k and $(k+1)$; (2) For a X node, that represents the mention j ends at k; or (3) for the both I_{k+1}^j and X nodes via a hyperedge that presents the co-occurrence of cases 1 and 2 at position k. A simple mention hypergraph model is then shown in Fig. 4.

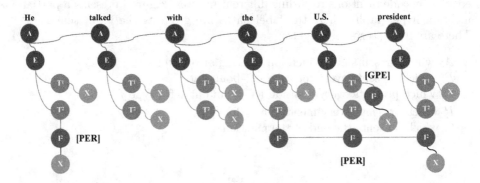

Fig. 4. A simple example illustrating the mention hypergraph model.

4.1   Encoded-Mention Hypergraph Model

A mention, i.e., j can be typically shown as a sequence of I^j nodes. The idea is to put on the encoding schema alongside more edges to broaden the feature set for a given hypergraph models. BI (beginning, inside) is the proposed encoded mention hypergraph model. For the BI encoded mention hypergraph model, node B alongside more edges span left-bound features of a given mention. Figure 5 illustrates one aspect of the proposed model, where every possible label paths for a given input sentences is given. The nodes can be categorized as either one of the following six types.

- A_k indicates all the mentions that are started at k or after k.
- E_k indicates all the mentions that the left boundary is at k position.
- T_k^j indicates all the mentions referring to the type j whose left boundary is at k position.
- B_k^j indicates all the mentions referring to the type j starting at k position.
- I_k^j indicates all the mentions referring to the type j covering k position.
- X indicates a mention ends at current position.

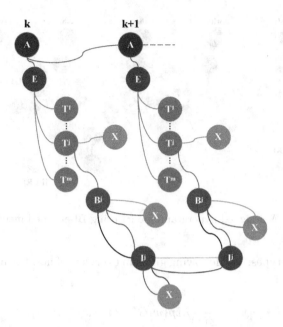

Fig. 5. A (partial) BI-encoded mention hypergraph model.

Figure 5 demonstrates the addition of the amount of connections for the edge. For each A_k node is coupled with an A_{k+1} and E_k nodes. Furthermore, for each E_k node is coupled with $(T_k^1, T_k^2, \ldots, T_k^m)$ using a hyperedge, where m denotes the number of textual mention types. This pair is identical to the pair in the mention hypergraph model. Every node T_k^j has a pair of edges which link to (1) a B_k^j node, where mention T^j is started at k position; or (2) for a X node, there are no mentions j are detected at k position. For each B_k^j node that is coupled with: (1) for an I_{k+1}^j node, we can have that mention j span k and $(k + 1)$; (2) for a X node, it shows a mention of base length at k position; or (3) for both I_{k+1}^j and X nodes, where both of the previous two cases exist at k position. For every I_k^j node possesses three edges such as: (1) for an I_{k+1}^j node, where a mention j spanning k and $(k + 1)$; (2) for an X node, where mention j ends at k; and (3) for both I_{k+1}^j and X nodes via a hyperedge, where both cases happen at k. Figure 6 given an illustration of a labeled sentence where the proposed hypergraph model using BI-encoded mention is identified.

4.2   Neural-Encoded-Based Mention Hypergraph Model

Neural networks were also utilized for the encoded hypergraph models. Bi-LSTM computes feature scores for both edges and hyperedges (if any). For a given (hyper-)edge at k, word parts of word k is input into the neural network. Using a linear/nonlinear transformation, the network finds the feature scores for the

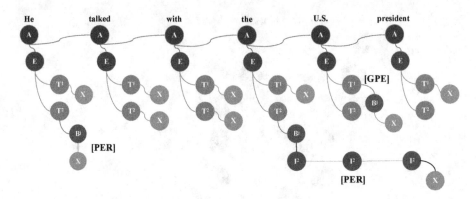

Fig. 6. A designed hypergraph model using BI-encoded mention.

various mention types. The following gives an overview of how the neural network is applied:

$$p(s|x) = \frac{1}{Z(x)} exp\{w_1 G(x, s) + w_2 N(x, s)\}, \qquad (4)$$

where $G(x, s)$ presents the feature scores. Furthermore, the w_1 and w_2 respectively denote the corresponding weights of the encoded features and neural features. For the $N(x, s)$, it shows the neural feature scores. In addition, the $Z(x)$ is the normalization factor of every label sequence over x. The maximum conditional likelihood estimation is used here for training the network. Assume that a training set is defined as $\{(x_i, s_i)\}$, we can then calculate the log-likelihood value as:

$$L_D(W) = \Sigma_{i \epsilon D} log \ p(s|x), \qquad (5)$$

in which W can thus be discovered to maximize the log-likelihood $L_D(W)$.

Initially, the proposed model forwards Bi-LSTM neural network. After that, the calculated feature scores are merged with the hypergraph spare features. The feature sets dictate the forward and backward passes both sparse and neural network feature). As the last step, the updated neural network features comprise the backward algorithm that is used to update the parameters in the neural network.

4.3   Features

The hypergraph features used to compute $G(x, s)$ are inspired by the model proposed by Finkel et al. [10]. In this work, we classify the features as the following:

- **Word features and POS tag features:** near the current position with a threshold of 3.
- **Word n-grams and POS n-gram features** carry the positions, i.e., $n = 2, 3, 4$.

- **Word-pattern features** involve several features, i.e., all-caps, all-digits, contains-hyphen, contains-dots, contains-digits, all-alphanumeric, lonely-initial, initial-caps, single-character, punctuation-mark, URL, and Roman-number.
- **Bag of words features** include the current word with a threshold of 5.

An additional feature as inspired by the model developed by Lu and Roth [17] is defined as:

- **Mention penalty:** The number of hyperedges used to link a pair of nodes T and B.

Through fine tuning the feature of mention penalty, the proposed architecture can thus learn the preference of mentions for any input length.

5   Experimental Results

Through experimentation and based on other research [17, 21], we performed experimentation on the **ACE2005** dataset. Table 1 shows the characteristics of the used **ACE2005** dataset. The number in parenthesis is the respective mention entity number. #Train, represent the number of sentences in the training set; #Dev shows the number of sentence in the development set; and #Test then indicates the number of testing set. The results are compared to multiple baseline approaches to show a comparison in terms of effectiveness for the of the designed approach. The hyperparameter max span length n has a significant impact on the performance of the semi-CRF model. Semi-CRF ($n = 6$) presents the model with $n = 6$. The remaining baseline approaches are identified with names and years. F in parenthesis represents the use of the mention penalty feature as explained as the additional feature explained in the previous section. Grid search approach is used to optimize this particular metric.

Table 1. Statistics of the used ACE2005 dataset.

	#Train	#Dev	#Test
ACE2005	7,336(24,687)	958(3,217)	1,047(3,027)

For this experimentation, the English portion of the used ACE2005 dataset is spitted following elder works. Besides, all documents from arabic treebank, chinese treebank, bnews, and nwire were used for this ACE 2005 dataset. Similar to other studies, 80% of the dataset is reserved as the training set, 10% for validation, and the last 10% of for testing the architecture. The bold fonts and underlines in Table 2 highlight the outperforming parts of the validation and test sets.

The presented results demonstrate that the proposed model achieves much better performance than the elder works. After the F-value optimization is utilized and performed [17], the F-measure value of the BI-NEMH shows the best performance among the compared approaches.

Table 2. The discovered results in the ACE2005 dataset.

	Dev set			Test set		
	Precision	Recall	F-measure	Precision	Recall	F-measure
Semi-CRF (n = 6)	≅75.3	≅48.5	≅59.0	≅72.8	≅45.0	≅55.6
Lu and Roth (2015)	≅79.3	≅50.6	≅61.8	≅76.9	≅47.7	≅58.9
Lu and Roth (2015) (F)	≅67.5	≅61.8	≅64.5	≅66.3	≅59.2	≅62.5
BI-NEMH	≅80.37	≅56.89	≅66.62	≅80.03	≅55.2	≅65.34
BI-NEMH(F)	≅72.89	≅65.12	≅68.79	≅73.09	≅63.36	≅**67.88**

6 Conclusion

This paper studies mention extraction and proposes a novel neural-encoded mention hypergraph (NEMH) as a candidate solution to address mention extraction tasks. The designed model can be used to utilize a BI-encoding schema that detects and records a richer feature set. The designed model also incorporates Bi-LSTM alongside the encoded hypergraph model, which facilitates the construction of the neural encoded mention hypergraph model. Extensive experimentation on three benchmark datasets gives an overview of the outperformance of the proposed model over state-of-the-art approaches and architectures. These results are expected to change the way researchers approach mention extraction tasks.

References

1. Berger, A.L., Pietra, S.A.D., Pietra, V.J.D.: A maximum entropy approach to natural language processing. Comput. Linguist. **22**(1), 39–71 (1996)
2. Baum, L.E., Petrie, T.: Statistical inference for probabilistic functions of finite state Markov chains. Ann. Math. Stat. **37**(6), 1554–1563 (1966)
3. Baum, L.E., Eagon, J.A.: An inequality with applications to statistical estimation for probabilistic functions of Markov processes and to a model for ecology. Bull. Am. Math. Soc. **37**(3), 360–363 (1967)
4. Baum, L.E., Sell, G.R.: Growth transformations for functions on manifolds. Pac. J. Math. **27**(2), 211–227 (1968)
5. Baum, L.E., Petrie, T., Soules, G., Weiss, N.: A maximization technique occurring in the statistical analysis of probabilistic functions of Markov chains. Ann. Math. Stat. **41**(1), 164–171 (1970)
6. Baum, L.E.: An inequality and associated maximization technique in statistical estimation of probabilistic functions of a Markov process. Inequalities **3**, 1–8 (1972)
7. Dyer, C., Ballesteros, M., Ling, W., Matthews, A., Smith, N.A.: Transition based dependency parsing with stack long short term memory. In: Conference on Association for Computational Linguistics, pp. 334–343 (2015)
8. Florian, R., et al.: A statistical model for multilingual entity detection and tracking. In: Annual Conference of the North American Chapter of the Association for Computational Linguistics, pp. 1–8 (2004)

9. Finkel, J.R., Manning, C.D.: Nested named entity recognition. In: Conference on Empirical Methods in Natural Language Processing, pp. 141–150 (2009)
10. Finkel, J.R., Kleeman, A., Manning, C.D.: Efficient, feature-based, conditional random field parsing. In: Conference on Association for Computational Linguistics, pp. 959–967 (2008)
11. Fine, S., Singer, Y., Tishby, N.: The hierarchical hidden Markov model: analysis and applications. Mach. Learn. **32**(1), 41–62 (1998)
12. Gupta, P., Andrassy, B.: Table filling multi-task recurrent neural network for joint entity and relation extraction. In: International Conference on Computational Linguistics, pp. 2537–2574 (2016)
13. Guo, S., Chang, M.W., Kiciman, E.: To link or not to link? A study on end-to-end Tweet entity linking. In: Conference on North American Chapter of the Association for Computational Linguistics, pp. 1020–1030 (2013)
14. Hochreiter, S., Schmidhuber, J.: Long short-term memory. Neural Comput. **9**(8), 1735–1780 (1997)
15. Huang, Z., Xu, W., Yu, K.: Bidirectional LSTM-CRF models for sequence tagging. https://arxiv.org/abs/1508.01991 (2015)
16. Lample, G., Ballesteros, M., Subramanian, S., Kawakami, K., Dyer, C.: Neural architectures for named entity recognition. https://arxiv.org/abs/1603.01360 (2016)
17. Lu, W., Roth, D.: Joint mention extraction and classification with mention hypergraphs. In: Conference on Empirical Methods in Natural Language Processing, pp. 857–867 (2015)
18. Lafferty, J.D., Mccallum, A., Pereira, F.C.N.: Conditional random fields: probabilistic models for segmenting and labeling sequence data. In: International Conference on Machine Learning, pp. 282–289 (2001)
19. McCallum, A., Freitag, D., Pereira, F.C.N.: Maximum entropy Markov models for information extraction and segmentation. In: International Conference on Machine Learning, pp. 591–598 (1999)
20. Mintz, M., Bills, S., Snow, R., Jurafsky, D.: Distant supervision for relation extraction without labeled data. In: Conference of ACL-ICJNLP, pp. 1003–1011 (2009)
21. Muis, A.O., Lu, W.: Labeling gaps between words: recognizing overlapping mentions with mention separators. In: Conference on Empirical Methods in Natural Language Processing, pp. 2598–2608 (2017)
22. Rosenberg, D.S., Dan, K., Taskar, B.: Mixture-of-parents maximum entropy Markov models. https://arxiv.org/abs/1206.5261 (2012)
23. Sarawagi, S., Cohen, W.W.: Semi-Markov conditional random fields for information extraction. In: Conference on Neural Information Processing Systems, pp. 1185–1192 (2004)
24. Zhuo, J., Cao, Y., Zhu, J., Zhang, B., Nie, Z.: Segment-level sequence modeling using gated recursive semi-Markov conditional random fields. In: Conference on Association for Computational Linguistics, pp. 1413–1423 (2016)

A Latent Variable CRF Model
for Labeling Prediction

Jerry Chun-Wei Lin[1], Jimmy Ming-Tai Wu[2(✉)], Yinan Shao[3], Matin Pirouz[4],
and Binbin Zhang[5,6]

[1] Department of Computer Science, Electrical Engineering and Mathematical
Sciences, Western Norway University of Applied Sciences, 5063 Bergen, Norway
`jerrylin@ieee.org`
[2] College of Computer Science and Engineering,
Shandong University of Science and Technology, Qingdao 266, China
`wmt@wmt35.idv.tw`
[3] School of Computer Science and Technology,
Harbin Institute of Technology (Shenzhen), Shenzhen 518055, China
`shaoyn0817@163.com`
[4] Department of Computer Science,
California State University, Fresno, CA 93740, USA
`mpirouz@ieee.org`
[5] Department of Biochemistry and Molecular Biology,
Shenzhen University Health Science Center, Shenzhen 518055, China
`zhangbb@szu.edu.cn`
[6] Center for Anti-aging and Regenerative Medicine,
Shenzhen University Health Science Center, Shenzhen 518055, China

Abstract. A latent variable conditional random fields (CRF) model is
proposed to improve sequence labeling, which utilizes the BIO encoding
schema as latent variable to capture the latent structure of hidden variables and observation data. The proposed model automatically selects
the best encoding schema for each given input sequence. Through experimentation, it is demonstrated that the proposed model unveils the latent
variable while performing robustly on sequence-labeling prediction tasks.

Keywords: Latent variable CRF · Sequence prediction · Encoding
schema · Natural language processing

1 Introduction

The first step of processing textual data is sequence prediction, where of
each unit/subsequence of input sequences is identified and semantic labels are
assigned. Sequence prediction facilitates the interpretation of the components
or structures of given contexts for machines. Conventionally, name entity recognition is used to perform this task, where name entities (i.e. person, company,
etc.) are captured from texts. Subsequently, chunking is used to detect the con-
stituent parts of sentences (i.e., nouns, verbs, adjectives, etc.) and information

© Springer Nature Singapore Pte Ltd. 2019
J. C.-W. Lin et al. (Eds.): MISNC 2019, CCIS 1131, pp. 68–78, 2019.
https://doi.org/10.1007/978-981-15-1758-7_6

extraction (i.e., author, title, journal, etc.) from a given reference string is done using reference parsing.

In this work, we utilize CRF model for sequence labeling. Sequence prediction has received substantial attention for a few decades, as it represents a fundamental research in the natural language processing. Sequence prediction plays an important role in downstream tasks, including the relation extraction [9,16], entity linking [8], and co-reference resolution [13]. Conventional sequence labeling models such as conditional random fields (CRF) and the maximum entropy model (MEM) provide the model of the conditional probability over the input sequence by representing the input unit, i.e. characters or words. The textual span of the input sequence, a.k.a the subsequence, is represented using segmentation models such as the semi-Markov random fields (semi-CRF). Various researchers [6,21] have demonstrated that the model performance is a characterized by the encoding schema. The BIO encoding schema is predominantly used for this purpose, where the beginning is shown as **B**, and the inside and outside are shown as **I** and **O**, respectively.

To explain BIO encoding schema, an example is provided in Fig. 1. The beginning of a person entity 'Michel' is labeled as B̲. Similarly, 'Bush' is also labeled as B̲ in BIO encoding schema but in contrast, 'Jordan' is labeled as I̲ as it is inside a person identity.

BIO encoding	Michel	Jordan	would	choose	Bush
	B-PER	I-PER	O	O	B-PER

Fig. 1. The BIO encoding schema.

The proposed model, based on a latent variable CRFs, selects the best encoding scheme for each given input sentence. The main contributions of the proposed model are outlined below:

- A latent variable CRFs model is developed for the task of sequence prediction, with the possibility of applicability to various sequence prediction subtasks, such as part-of-speech tagging, name entity recognition, chunking, etc.
- The proposed model incorporates encoding schema to extract the hidden structures of the hidden variables and observation data. Thus, automatic selection of the best encoding schema for the whole input sentence is possible.
- Experimentation demonstrates that the proposed latent variable model outperforms the conventional CRF with BIO and BILOU encoding schemas.

2 Literature Review

Traditional mention-extraction models include linear models based on the hidden Markov model (HMM) [1–3], max-entropy model (MEM) [4], conditional random fields (CRF) [11], and semi-Markov random fields (semi-CRF) [25]. These are linear models that can capture correlations between labels in neighborhoods and jointly decode the best chain of labels, given an input sequence. Leonard and coworkers [1–3] proposed a hidden Markov model (HMM) that can be represented as a dynamic Bayesian network. Fine et al. [7] proposed a hierarchical hidden Markov model (HHMM), which is a recursive hierarchical generalization of the vanilla hidden Markov model. Zhang et al. [29] built the ICTCLAS system, which uses a hierarchical hidden Markov model to incorporate Chinese word segmentation, part-of-speech tagging, disambiguation, and unknown-word recognition in a comprehensive theoretical frame. Berge et al. [4] pioneered a maximum entropy model (MEM) in natural language processing. McCallum et al. [15] proposed a maximum entropy Markov model (MEMM), which is a graphical model for sequence prediction that combines features of HMM and MEM. Ratnaparkhi [22] proposed a statistical model that trains from a corpus annotated with part-of-speech tags and assigns them to previously unseen text with high accuracy. Rosenberg et al. [24] then proposed the mixture-of-parents maximum entropy Markov model (MoP-MEMM). Which allows tractable incorporation of long-range dependencies between nodes by restricting the conditional distribution of each node to a mixture of parent distributions.

Conditional random field (CRF) models were proposed by Lafferty et al. [11], which is a class of statistical modeling method often applied in sequence-prediction problems. Tseng et al. [28] presented a Chinese word segmentation (CWS) system based on CRF models. Zhao et al. [30] regarded the CWS problem as a character-based tagging problem under a conditional random fields framework. Cuong et al. [5] considered the problem of incorporating high-order dependencies between labels or segments in conditional random fields. Semi-Markov conditional random fields (semi-CRF) were proposed by Sarawagi and Cohen [25]. Importantly, features of semi-CRFs can measure properties of segments, and transitions within a segment can be non-Markovian. Okanohara et al. [19] presented techniques to apply semi-CRFs to named entity-recognition tasks with a tractable computational cost. Nguyen et al. [18] extended first-order semi-Markov semi-CRFs to include higher-order semi-Markov features, and presented efficient inference and learning algorithms, under the assumption that the higher-order semi-Markov features are sparse. Muis and Wei [17] proposed weak semi-Markov conditional random fields for noun-phrase chunking. In conventional semi-CRF, the model intuitively decides the next segment length and type at the same time, while in a weak semi-CRF, the model tries to propose a weaker variant that makes the two decisions separately by restricting each node to connect to either only the nodes of the same label in the next segment, or to all nodes only in the next word. The weak semi-CRF model yields performance similar to that of conventional semi-CRFs, but runs significantly faster.

Researchers have also applied deep learning methods on sequence labeling with promising results. Huang et al. [10] proposed a variety of architectures for detecting and predicting sequences using long short-term memory (LSTM), bidirectional LSTM (Bi-LSTM) networks, LSTM with a CRF layer (LSTM-CRF), and bidirectional LSTM with a CRF layer (Bi-LSTM-CRF).

A neural semi-Markov conditional random field model was developed by Liu et al. [12], which comprises both the embedding of both input units and segments. Their experimentation with name entity recognition (NER) and Chinese word segmentation (CWS) tasks showed successful results. Ma and Hovy [14] proposed a CNN-LSTM-CRF model with word- and character-level representations. Through an end-to-end model, their proposed architecture did not require feature engineering or data pre-processing. Rei et al. [23] incorporated character-level information to handle the out-of-vocabulary (OOV) issue in sequence prediction. They investigated character-level extensions to conventional LSTM-CRF structure models. The encoded character-level information was combined with pre-trained word embeddings using an attention mechanism, which dynamically decides how much information to use from a word- or character-level component.

The sequence prediction has been previously achieved by different approaches based on latent variable models. Sun and Nan [27] proposed a latent discriminative model, called Latent Semi-CRF, which incorporates advantages of two modeling approaches Petrov and Klein [20] reported a discriminative latent variable approach for syntactic parsing in which rules exist at multiple scales of the refinement. Sun et al. [26] proposed a latent semi-CRF model to synchronously detect the new words together with their POS regardless of the type of the new words from the Chinese text without being pre-segmented.

3   Preliminaries and Problem Statement

This section briefly introduces the preliminaries and problem statements of our works.

3.1   Latent Variable CRF

Consider a sequence of observations $x = (x_1, \ldots, x_n)$. In the latent variable CRF, the model must determine how to assign a sequence of labels $y = (y_1, \ldots, y_n)$, from one finite set of labels Y. Instead of directly modeling $P(y|x)$, as a conventional CRF would do, a set of latent variables h is "inserted" between the x and y using the chain rule of probability, i.e.,

$$P(y|x) = \frac{1}{Z(x)} \sum_h P(y|h, x)P(h|x), \tag{1}$$

where $Z(x)$ is the normalization factor, h denotes the latent variable, x is the sequence of observations, and y represents the sequence of labels. This model allows capturing latent structure between the observations and labels. These models find applications in computer vision, specifically gesture recognition from video streams and sequence prediction.

3.2   Encoding Schema

BIO encoding represents the most popular encoding schema. Figure 2 shows an example of BIO encoding schema, where **B** denotes the beginning of a segment, **I** represents the inside (including the ending) of a segment, and **O** stands for the word which does not belong to any segment. As shown in Fig. 2, 'Michel' is the begin word of a person entity, thus, it is marked with **B-P** (Begin-Person), 'Jordan' is the inside word of a person entity, thus, it is marked with **I-P** (Inside-Person). For word 'would', since it does not belong to any entity, it is marked with **O**.

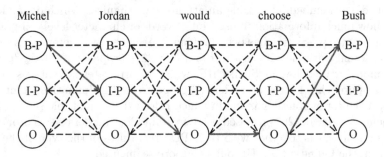

Fig. 2. An example of BIO encoding schema.

3.3   Problem Statement

Formally, considering an input sequence $x = (x_1, \ldots, x_k)$ of length k, a label of x is defined as a tuple (u, y) which means the u-th input word is associated with label y. A label sequence of x is then defined as $s = (s1, \ldots, sk)$, where $s_j = (u_j, y_j)$. It should be noted that the input sequence x and the label sequence s have equal lengths. Given an input sequence x, the sequence prediction problem is defined as the problem of finding the most probable label sequence s of x.

4   The Proposed Latent Variable CRF Model

This section introduces the proposed latent variable CRF model. We briefly introduce the conventional CRF model, then state the proposed latent variable CRF model. The proposed latent variable CRF is a sentence level model, which can automatically determine the best encoding schema for sequence prediction. Details are respectively described as below.

4.1   Conventional CRF

Conditional random field (CRF) is a popular model for sequence labeling task. Compared with other models such as hidden Markov model or maximum entropy models (MEM), CRF can easily incorporate flexible features and handle the label bias problem in MEM model. Figure 3 shows architecture of the conventional CRF without using any encoding schema, P node denotes the person name entity node and O node denotes the non-entity node. In Fig. 3, the dashed lines encode all the possible labeled path of the given input sequences. Since the supervised training is utilized in the designed model, there always has a labeled path (i.e., the red line) in CRF model, which corresponds to the given label. During the training procedure, the model will tune the parameters to maximize the probability of this labeled path. The CRF model is the conditional probability of the possible out sequence s over the input sequence x, as:

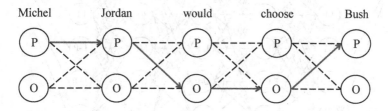

Fig. 3. Conventional CRF. (Color figure online)

$$p(s|x) = \frac{1}{Z(x)} exp\{W \cdot G(x, s)\}, \tag{2}$$

where $G(x, s)$ is the feature function, W is the weight vectors, and $Z(x)$ is the normalization factor. To find the best label sequence in the CRF, let σ_j denote the best label sequence ends of the j-th input, (m, n, y) denote a label sequence start at the m-th position, end at nth position and labeled as y. Next, σ_j is recursively calculated as:

$$\sigma_j = max\Psi(j - 1, j, y) + \sigma_{j-1}, \tag{3}$$

where $\Psi(j - 1, j, y)$ is the feature value defined over the label sequence $s = (j - l, j, y)$.

4.2 Proposed Latent Variable CRF Model

Compared with conventional CRF, our proposed model incorporate hidden variables to explore more information in input sequence. Figure 4 shows architecture of the first proposed latent variable CRF.

The proposed graph model is formed as follows. It corresponds to the CRF with BIO encoding schema in Fig. 4. The connection relation is as follows. The **B** node may be connected to **I**, **O** and **B** nodes. The **I** node may be connected to **I**, **O** and **B** nodes. The **O** nodes may be connected to **B** and **O** nodes, while it cannot be connected to an **I** node, since the begging of a segment must be labeled with **B** node.

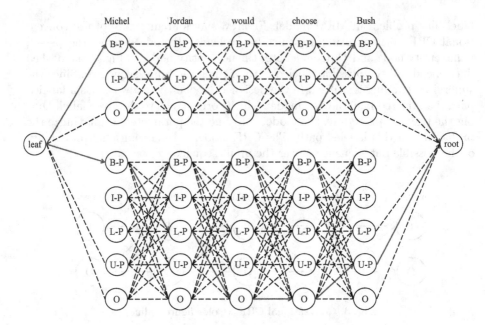

Fig. 4. Proposed latent variable CRF model.

As shown in Fig. 4, for a given input sentence, the proposed graph model provides one labeled paths, i.e., it corresponds to the BIO encoding schema. We consider the encoding schema as the latent variable. In the decoding step, the model will use Viterbi algorithm to choose one of the red paths with the highest feature scores as the final output label sequence. Through this framework, for a given input sentences, the model is forced to automatically determine a encoding schema. This is the main difference from the conventional CRF models. For conventional CRF model in Fig. 3, there is only one labeled path in the model corresponding to one certain encoding schema or no encoding schema. But this only uses part of the information, since one sentence can be represented by different encoding schema. The proposed model uses the information of two labeled path, thus the model itself can determine and choose the suitable encoding schema.

4.3 Training Procedure

Following the CRF, we adopted a log-linear approach for such a latent variable CRF. Specifically, for the given input sentence x, the probability of predicting a possible output sequence y reads:

$$p(y|x) = \frac{exp(w^T f(x, y))}{\sum_{y'} exp(w^T f(x, y'))},$$ (4)

where $f(x, y)$ is the feature vector defined over the input-output pair (x, y), and the weight vector w gives the parameters of the model. Our objective is to minimize the regularized negative joint log-likelihood of the dataset, as:

$$L(w) = \sum_i log \sum_{y'} exp(w^T f(x_i, y')) - \sum_i w^T f(x_i, y) + \lambda w^T w, \qquad (5)$$

where (x_i, y_i) refers to the i-th training instance, and the last term is a $L2$ regularization term with λ setting to 0.01. This objective function may be optimized with standard gradient-based methods. We use the L-BFGS algorithm as our optimization method and Viterbi algorithm as our inference method in this work.

4.4   Features

Let us we briefly introduce the CRF feature to compute the $\mathbf{G(x,s)}$ in Eq. 2. Specifically, we consider the following features defined over the inputs.

- **Words features:** Words that appear around the current position with a window of size 3.
- **POS tags features (if available):** POS tags that appear around the current position with a window of size 3.
- **Word n-grams features:** Word n-grams that contain the position, for $n = 2, 3, 4$.
- **POS n-grams features (if available):** POS tags that contain the current position, for $n = 2, 3, 4$.

All these features are used in all conventional CRF based models and the proposed latent variable model for comparisons. It worth to mention that the features used in this are quite simple but still show good performance in terms of accuracy, since the emphasis in this work is the effectiveness of the proposed framework rather than feature engineering. The motivation of this paper is to present a framework without task specific feature engineering. Of course, with task specific feature engineering, the performance in terms of accuracy of the designed models can be enhanced.

5   Experimental Evaluation

In this section, we evaluate our models on three natural language processing tasks, i.e., the name entity recognition and chunking. We thoroughly compare the performance of the proposed latent variable CRF with conventional CRF with BIO and BILOU encoding schemas, respectively. The model uses the same feature described in Sect. 4.4. The CRF-BIO is the conventional CRF with BIO encoding schema, and the CRF-BILOU is the conventional CRF with BILOU encoding schema. The designed model is then evaluated as follows.

5.1 Data Sets

In this section, we use the standard datasets to evaluate the performance of the designed models and the compared ones. Table 1 lists the corpora statistics of the conducted datasets. Details are described as follows.

Table 1. Corpora statistics of the used datasets.

Name	Task	# labels	# train	# dev	# test
CoNLL2003	NER	8	14,987	3,466	3,684
CoNLL2000	Chunking	22	8,936	N/A	2,012

5.2 Name Entity Recognition (NER)

Table 2 compares the performance of different models, where the best performance is marked with underline. The designed model may be viewed as a combination of the CRF-BIO, where its performance is robust and outperforms both the CRF-BIO and CRF-BILOU. This explains the fact that the designed model may automatically choose the best encoding schema for the input sentence. From experiments, it can be shown that, for a give input sentence, the designed model can effectively choose best encoding schema for the sentence. As expected, the CRF-BIO and CRF-BILOU exhibit a poor performance, as the performance of the CRF with BIO encoding schema.

Table 2. Results on the Conll2003 dataset.

NER task	Precision	Recall	F1
CRF-BIO	84.10	83.59	83.84
CRF-BILOU	83.82	84.36	84.09
Proposed model	**84.19**	**84.71**	**84.46**

5.3 Chunking

Table 3 compares the performance of different models for the chunking task on CoNLL2000 shared task. It may be seen from this table that, the proposed model outperforms the baseline models, CRF-BIO and CRF-BILOU. The CRF with BIO encoding schema performs better in the chunking task, while the CRF with BILOU encoding schema outperforms in the name entity recognition. This is due to the fact that none of the encoding schemas is the best, it is necessary to choose different encoding schemas for different input sentences as we do in this work.

Table 3. Results on the CoNLL2000 dataset.

Chunking task	Precision	Recall	F1
CRF-BIO	**90.15**	89.89	90.01
CRF-BILOU	90.05	89.88	89.96
Proposed model	90.12	**90.23**	**90.17**

6   Conclusion

This paper focuses on the task of sequence-labeling prediction. The proposed model is capable of choosing the best encoding schema for a given input sentence. Empirically, we have demonstrated the effectiveness of the proposed model across several standard sequence prediction tasks and datasets.

References

1. Baum, L.E.: An inequality and associated maximization technique in statistical estimation of probabilistic functions of a markov process. Inequalities **3**, 1–8 (1972)
2. Baum, L.E., Eagon, J.A.: An inequality with applications to statistical estimation for probabilistic functions of markov processes and to a model for ecology. Bull. Am. Math. Soc. **37**(3), 360–363 (1967)
3. Baum, L.E., Petrie, T.: Statistical inference for probabilistic functions of finite state Markov chains. Ann. Math. Stat. **37**(6), 1554–1563 (1966)
4. Berger, A.L., Pietra, S.A.D., Pietra, V.J.D.: A maximum entropy approach to natural language processing. Comput. Linguist. **22**(1), 39–71 (1996)
5. Cuong, N.V., Ye, N., Lee, W.S., Chieu, H.L.: Conditional random field with high-order dependencies for sequence labeling and segmentation. J. Mach. Learn. Res. **15**(1), 981–1009 (2014)
6. Dai, H., Lai, P., Chang, Y., Tsa, R.T.: Enhancing of chemical compound and drug name recognition using representative tag scheme and fine-grained tokenization. J. Cheminformatics **7**(1), 1–10 (2015)
7. Fine, S., Singer, Y., Tishby, N.: The hierarchical hidden Markov model: analysis and applications. Mach. Learn. **32**(1), 41–62 (1998)
8. Guo, S., Chang, M.W., Kiciman, E.: To link or not to link? a study on end-to-end tweet entity linking. In: The Conference of the North American Chapter of the Association of Computational Linguistics, pp. 1020–1030 (2013)
9. Gupta , P., Andrassy, B.: Table filling multi-task recurrent neural network for joint entity and relation extraction. In: The International Conference on Computational Linguistics, pp. 2537–2547 (2016)
10. Huang, Z., Xu, W., Yu, K.: Bidirectional LSTM-CRF models for sequence tagging (2015). http://arxiv.org/abs/1508.01991s
11. Lafferty, J.D., Mccallum, A., Pereira, F.C.N.: Conditional random fields: probabilistic models for segmenting and labeling sequence data. In: The Eighteenth International Conference on Machine Learning, pp. 282–289 (2001)
12. Liu, Y., Che, W., Guo, J., Bin, Q., Liu, T.: Exploring segment representations for neural segmentation models. In: The International Joint Conference on Artificial Intelligence, pp. 2880–288 (2016)

13. Lu, J., Venugopal, D., Gogate, V., Ng, V.: Joint inference for event coreference resolution. In: The International Conference on Computational Linguistics, pp. 3264–3275 (2016)
14. Ma, X., Hovy, E.: End-to-end sequence labeling via bi-directional LSTM-CNNS-CRF. In: The Annual Meeting of the Association for Computational Linguistics, pp. 1064–1074 (2016)
15. McCallum, A., Freitag, D., Pereira, F.C.N.: Maximum entropy Markov models for information extraction and segmentation. In: The International Conference on Machine Learning, pp. 591–598 (1999)
16. Mintz, M., Bills, R.S.S., Jurafsky, D.: Distant supervision for relation extraction without labeled data. In: The Annual Meeting of the Association for Computational Linguistics, pp. 1003–1011 (2009)
17. Muis, A.O., Lu, W.: Weak semi-Markov CRFS for noun phrase chunking in informal text. In: The North American Chapter of the Association for Computational Linguistics: Human Language Technologies, pp. 714–719 (2016)
18. Nguyen, V.C., Lee, W.S., Ye, N., Hai, L.C.: Semi-Markov conditional random field with high-order feature, pp. 1–4 (2011)
19. Okanohara, D., Miyao, Y., Tsuruoka, Y., Tisuji, J.: Improving the scalability of semi-Markov conditional random fields for named entity recognition. In: The Annual Meeting of the Association for Computational Linguistics, pp. 465–472 (2006)
20. Petrov, S., Dan, K.: Sparse multi-scale grammars for discriminative latent variable parsing. In: The Conference on Empirical Methods in Natural Language Processing, pp. 867–876 (2008)
21. Ratinov, L., Roth, D.: Design challenges and misconceptions in named entity recognition. In: The Conference on Computational Natural Language Learning, pp. 147–155 (2009)
22. Ratnaparkhi, A.: A maximum entropy model for part-of-speech tagging. In: The Conference on Empirical Methods in Natural Language Processing, pp. 133–142 (1996)
23. Rei, M., Crichton, G.K.O., Pyysalo, S.: Attending to characters in neural sequence labeling models (2016). http://arxiv.org/abs/1611.04361
24. Rosenberg, D.S., Dan, K., Taskar, B.: Mixture-of-parents maximum entropy Markov models (2012). http://arxiv.org/abs/1206.5261
25. Sarawagi, S., Cohen, W.W.: Semi-Markov conditional random fields for information extraction. In: The Neural Information Processing Systems, pp. 1185–1192 (2004)
26. Sun, X., Huang, D., Ren, F.: Detecting new words from chinese text using latent semi-CRF models. IEICE Trans. Inform. Syst. **93**(6), 1386–1393 (2010)
27. Sun, X., Nan, X.: Chinese base phrases chunking based on latent semi-CRF mode. In: The International Conference on Natural Language Processing and Knowledge Engineering, pp. 1–7 (2010)
28. Tseng, H., Chang, P., Andrew, G., Jurafsky, D., Manning, C.: Sequential labeling with latent variables. In: The Workshop on Chinese Language Processing, pp. 168–171 (2015)
29. Zhang, H.P., Liu, Q., Cheng, X.Q., Zhang, H., Yu, H.K.: Chinese lexical analysis using hierarchical hidden Markov model. In: The Workshop on Chinese Language Processing, pp. 63–70 (2003)
30. Zhao, H., Huang, C.N., Li, M., Kudo, T.: An improved Chinese word segmentation system with conditional random field. In: The Workshop on Chinese Language Processing, pp. 162–165 (2006)

Exploring the Influencing Factors of Live-Streaming Viewers' Participation Intention from the Perspective of Source Credibility Model and Cognitive Load - An Example of Mobile Device Users

Shu-Chen Yang, Tzu-Ting Feng, and Yu-Hui Wang[✉]

Department of Information Management, National University of Kaohsiung,
Kaohsiung, Taiwan
henryyang@nuk.edu.tw, s911297@gmail.com,
well147258well@gmail.com

Abstract. In this study, we divide various behaviors of live-streaming viewers into non-creating, contributing and money-support three levels to exploring the influencing factors of live-streaming viewers' participation intention. Streamers being the source of information in the virtual live-streaming room, their traits or the messages they transmitted deeply affect the viewers' psychological state and behaviors that the viewers may act on the live-streaming platform. In addition, since the interactive mode of live-streaming program is that the streamer needs to face a large number of viewers in the same time, which means there is almost no direct interaction between the streamer and the viewers. The viewers only mostly imagine the interactive relationship between the two parties. In summary, we investigate the interaction of source credibility model, para-social interaction, cognitive load and the consequent outcome of participation intention for live-streaming platforms.

Keywords: Source credibility model · Para-social interaction · Participation intention · Cognitive load

1 Introduction

1.1 Research Background

Since Amazon acquired game streaming platform Twitch with 970 million dollars in 2014, Twitter acquired civilian streaming platform Periscope with 10 million dollars in 2015, and the addition with Facebook introducing Facebook Live, Google announcing YouTube Gaming. All these involvements with major tech companies make people start to realize the concept of online streaming, while indirectly reducing promotional expenses for later joining businesses.

© Springer Nature Singapore Pte Ltd. 2019
J. C.-W. Lin et al. (Eds.): MISNC 2019, CCIS 1131, pp. 79–92, 2019.
https://doi.org/10.1007/978-981-15-1758-7_7

1.2 Research Motivation and Purpose

After viewers pay for the said functions, they can have a "more outstanding trait" to differentiate themselves from others in live-stream chats. This way of making viewers willing to pay through streamers has already become a very important business model in live-streaming industries. Because of this, this research would like to understand how the interactive relationship between streamers and viewers might influence viewers' behavioral intentions, also providing live-stream business owners suggestions and references.

Hu et al. (2017) mentions that, current research fields about live streaming mostly discuss how to upgrade communication quality of live-streaming services through information technologies (e.g. Siekkinen et al. 2018). In addition, a few of them discuss live-stream viewers' behaviors (e.g. Hilvert-Bruce et al. 2018). Within the researches about live-stream viewers' behaviors, Wohn et al. (2018) discussed such behaviors from two aspects: intangible support and tangible support. Continuing with said research's concepts, this research also considers that not only financial supports, but also nonfinancial supports are also an important way to express supports towards streamers. Therefore, this research considers that it is necessary to discuss viewers' participation intentions towards streamers on platforms from both angles of nonfinancial supports and financial supports.

Wohn et al. (2018) separated the interaction model from streamers to viewers into two elements: broadcast element and interpersonal element. Facing a massive number of viewers, streamers can only briefly browse viewers' comments and make short replies. Such interaction model also makes viewers have an illusion of interacting with streamers but not exactly like chatting with friends in real life. Moreover, this concept of human online interaction is para-social interaction proposed by Horton and Wohl (1956).

The reason that live-stream viewers will have such para-social interactions with streamers is that streamers' performances during streaming successfully attract viewers, making viewers want to interact with them. Therefore, understanding which characteristics streamers have can make viewers have para-social interactions is quite important. Source credibility model by Ohanian (1990) includes three variables: expertise, trustworthiness, and attractiveness. They are used to reflect how viewers feel towards media characters' positive characteristics. This research considers that these three variables match the characteristics of streamers when they perform during live-streams. Therefore, it views source credibility model as a key factor on the side of streamers.

Lee (2017) considered that cognitive load, error resolution, and visual aesthetics are all key factors that influences operational experiences on interfaces like computers and mobile phones. And the size of mobile devices may influence how information presents, thus reducing viewers' cognitive loads when they watch live-streams or operate their devices can make their viewing experiences better. Lee (2017) also mentioned that cognitive load has greater influences towards mobile app users than factors influencing platforms like effectiveness, efficiency, and satisfaction. Therefore, this research considers whether cognitive load has a moderating effect on the relationship between para-social interaction experience and participation intention of viewers towards streamers or not is worth a discussion. This is why cognitive load is viewed as a key factor on the side of live-stream platforms.

As mentioned above, on both theoretical and practical aspects, this research has a certain level of distributions. First, since live-stream is counted as a relatively new industry in recent years, researches on it are not full and wide as researches on other social medias like social network sites. Through this research's discussion on live-stream viewers' participation behaviors, the application of para-social interaction and source credibility model on the field of live-stream can be expanded. In addition, this research further discusses influences towards streamers and viewers from factors on the side of live-stream platforms, using complete theories from cognitive load related field. Finally, the result of this research can be provided to live-stream related business owners for a better understanding of the interactive ecosystem between viewers and streamers, making itself a practical reference.

2 Literature Review

2.1 Live-Streaming and Participation Intention

In early literatures about live streaming, most of them discussed the appearance of live-stream platforms that made huge virtual online communities among online players (e.g. Smith et al. 2013). After that, there were researchers starting to discuss the influence from interactive relationships between streamers and viewers towards viewers' afterwards behaviors. For instance, Chen and Lin (2017) used the concepts of flow theory and social interaction to explain viewers' intentions of continuous viewing.

Following the evolution of live-stream techniques, interactions between viewers and media characters also appear to have some changes. For an example, viewers can gain streamers' attentions by giving them virtual gifts. And according to consumers' online brand-related activities (COBRAs) made by Muntinga et al. (2011), they divided consumers' behaviors of brand related participation on social medias into three levels: consuming, contributing, and creating. This research considers this classification can be applied to discuss modes of viewers' behaviors.

First, when it comes to a live-stream scenario, non-creating participations refer to viewers may observe what streamers say and do, or express their support towards streamers using only simple functions. Second, contributing participations (contributing levels) mean under the circumstances of live-streaming, viewers voluntarily make contributing behaviors for streamers through commenting with streamers or sharing links to streamers' live-stream pages. And the third level is creating level. To the live-streaming circumstances, it is more like when viewers voluntarily post online contents like taking photos with streamers. However, since viewers' creating behaviors are usually performed on social medias outside of the live-stream platforms, which do not belong to what this research is about, the creative level will not be discussed afterwards.

Wohn et al. (2018) used the concepts of emotional support and tangible support from social support theory to discuss what viewers' motivations to support streamers are. Emotional support is to express supports to streamers through verbal cares; while tangible support refers to express supports to streamers through offering services, helps, or money. This research will combine the consuming and contributing levels from the research of Muntinga et al. (2011), and the concept of financial support from Wohn et al. (2018) to discuss viewers' three levels of participation in showing supports from actions to finances, while using the concept of para-social interaction to explain the following hypothesis proposition.

2.2 Para-Social Interaction (PSI)

The terms "para-social interaction" can be traced back to the proposition from Horton and Wohl (1956). It describes a social relationship happening between media characters and viewers. Viewers of television may put themselves into scenarios of television shows they watch, and only from their prospects, interacting with media characters. Horton and Wohl described this kind of virtual interpersonal interaction as para-social interaction.

Rubin and McHugh (1987) explained para-social interaction as a "one-sided" interpersonal interactive relationship from viewers towards media characters. And Horton and Wohl (1956) mentioned that manipulating media characters could be helpful to the making of viewers' para-social interactions. This kind of relationship between media characters and viewers seems like interactions, however, without an actual interactive mode, is what para-social interactions are based on.

As mentioned above, under the circumstance of para-social interaction, actual interaction and beneficial behavior almost does not exist between media characters and viewers. The interactive relationships are from viewers' imaginations. And when viewers develop a one-sided, durable, intimate relationship through frequent contacts with media characters, this can be counted as a para-social relationship (Dibble et al. 2016). This kind of relationship always exist only in viewers' minds.

Para-Social Interactions in Online Environment

Following the popularity of the Internet, more researchers applied para-social interactions onto interactive medias based on the Internet (e.g. Gong and Li 2017). These researches not only assured the concept of para-social interaction not being limited to traditional medias, but also make viewers feel closer to media characters through the development of online environment.

For instance, the research of Pan and Zeng (2018), which discussed the skin colors of athletes and spectators, has found out that spectators express higher levels of para-social interaction experiences to athletes of the same skin colors. Furthermore, Gong and Li (2017) discussed the effects of celebrities endorsing products on micro blogs. Their result indicated that the para-social interaction experiences formed between fans and celebrities not only affect fans' trust levels towards celebrities, but also affect fans' attitudes towards commercial endorsements of celebrities.

Combing with what mentioned above, there are researchers in many fields discussing the interactive relationships between viewers and media characters and following business values, using para-social interaction as a variable. As a new social media, online live-stream allows viewers and streamers to have audiovisual experiences from immediate interactions, which can never be achieved by past traditional medias and recent digital medias. Based on this, this research will discuss the phe-nomenon of viewers' para-social interactions on live-stream platforms.

Online live-stream not only provides viewers immediate visual experiences, but also allows viewers to have immediate message communications with streamers (Hu et al. 2017). For an example, under normal live-stream circumstances, the number of viewers is bigger than streamers, which makes streamers not able to reply all com-ments. Therefore, in most circumstances, whatever streamers do, like greeting, having small chats, can merely be a "business" performance, but not an interaction with specific viewers. However, these behaviors can be read by viewers as if streamers are greeting them or sharing feelings with them. This illusion of two-sided interactions is known as a para-social interactive experience under the circumstance of online live-streams.

When viewers feel concerns towards themselves from streamers, they may want to interact with streamers, or having ideas of supporting streamers. Therefore, viewers may turn from following silently into supporting with actions, even through financing to support streamers' idea and thoughts, or in order to continuously see contents from streamers for future days (Wohn et al. 2018). Thus, this research will make a propo-sition of Hypothesis 1 to 3 as below:

H1: Viewers' para-social interactive experiences towards streamers may positively influence their non-creating participation intentions.

H2: Viewers' para-social interactive experiences towards streamers may positively influence their contributing participation intentions.

H3: Viewers' para-social interactive experiences towards streamers may positively influence their financial participation intentions.

2.3 Source Credibility Model

Source credibility model refers to the characteristic of reliability perceived by con-sumers about a source of messages. It influences the level of receivers' acceptances to messages (Ohanian 1990). For instance, I feel like a certain endorser is professional and trustworthy. It mainly focuses on the level of perception (e.g. Amos et al. 2008); Source credibility model is when consumers perceive their connection with messengers getting stronger, the more likes and approvals from them towards messengers. For an example, I feel that a certain endorser seems familiar, intimate, or similar to myself. It focuses on the level of relationship (e.g. Erdogan 1999).

Ohanian (1990) found out that the academic field at that time had inconsistent standards of source credibility measurement. Thus, they decided to simplify source attractiveness model into one single variable of attraction and merge it into source credibility model. In other words, when compared to the version of Hovland et al. (1953), Ohanian's (1990) source credibility model includes these three variables: expertise, trustworthiness, and visual attractiveness. And this version became a refer-ence for many following researchers to discuss source credibility.

In this research, while online live-stream being a kind of online community, and streamers being messengers, their credibility must influence viewers' attitude towards them. Therefore, this research will use Ohanian's (1990) credibility model to discuss streamers' positive characteristics and propose various details of variables and hypothesis as below:

i. Expertise

Ohanian (1990) considered expertise as an ability, power, and qualification of messengers. And the research of Xiang et al. (2016), which discussed consumers' behaviors on social commercial platforms, mentioned that when consumers are able to provide some professional shopping advices on platforms, it becomes possible that consumers want to interact and communicate with these shopping experts. In other words, expertise of sources can be one reason that makes community members want to interact with the sources of messages.

ii. Trustworthiness

Ohanian (1990) explained trustworthiness as the trusts and acceptances from receivers towards the messages sent by messengers. Singh and Sidhu (2015) mentioned that, as a base of any relationship, trust plays a very important role on maintaining relationships between social media members. Therefore, this research considers that the trustworthiness of messengers is also one of the factors, which make community members want to interact with them.

iii. Attractiveness

To avoid any confusion on definition, the term "attractiveness" in this research only refers to whether viewers consider streamers being attractive or not. Merkle and Richardson (2000) indicated that visual attractiveness constructs out images towards strangers, and being a reference for us to choose whether to keep communicating with them. Therefore, this research considers that visual attractiveness being one possible factor, which makes viewers want to interact with messengers.

As mentioned above, under the circumstance of online live-streams, expertise represents viewers' thoughts on whether streamers are familiar with discussed contents or not; trustworthiness represents viewers' perceptions of how charming or appreciated streamers' exteriors are. Therefore, this research considers that when viewers perceive more streamers' characteristic of credibility, viewers are more likely to immerse into the atmospheres created by streamers. Thus, viewers may see streamers as their companions and view streamers' behaviors as interactions with themselves. Therefore, here are Hypothesis 4 to 6 of this research's proposition:

H4: Streamers' expertise perceived by viewers may positively influence viewers' para-social interactive experiences with streamers.

H5: Streamers' trustworthiness perceived by viewers may positively influence viewers' para-social interactive experiences with streamers.

H6: Streamers' visual attractiveness perceived by viewers may positively influence viewers' para-social interactive experiences with streamers.

2.4 Cognitive Load

Lee (2017), after sorting prior user-interface-related literatures, has found out that many constructs like learnability, effectiveness were all very important to normal window interfaces and mobile application interfaces. However, since screen sizes of mobile application interfaces are comparatively smaller, making cognitive load become the only construct which mobile application interfaces must consider while normal window interfaces do not.

Lahtinen's (2017) research of visualizing mobile application interfaces has mentioned that when it came to designing mobile application interfaces, it was necessary to reduce users' cognitive loads through appropriate zooming functions, elements of interactions and avoiding over-messy compositions. To the circumstance of live-stream, when interfaces being designed, live-stream platform should be made with clean compositions to reduce viewers' cognitive loads. Under that condition, viewers may more possibly immerser into interactions with streamers.

This research uses references from Madden et al. (1992) to compare thesis from theory of reasoned action and theory of planned behavior. To the circumstance of live-stream, viewers perceive under higher pressures from systems (high cognitive loads) when they use platforms or watch live-streams with badly designed interfaces. And that may reduce viewers' perceived behavior controls. Under the circumstance, even if viewers have great para-social interactive experiences or attitudes towards streamers, it is possible to not increase viewers' willingness to support streamers. Therefore, this research makes this proposition:

H7: Cognitive load may negatively moderate the relationship between para-social interactive experience and non-creating participation intention.
H8: Cognitive load may negatively moderate the relationship between para-social interactive experience and contributing participation intention.
H9: Cognitive load may negatively moderate the relationship between para-social interactive experience and financial participation intention.

So our research has the research model as below (Fig. 1):

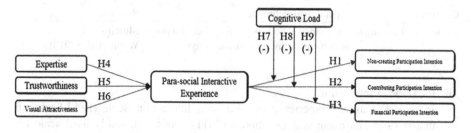

Fig. 1. Research Model for research

3 Research Methodology

3.1 Questionnaire Design

The research targets of this research are viewers who have previous experiences of watching online live-streams. This research then discusses their participation intentions on live-stream platforms, and using the snowball technique to issue questionnaires on Facebook. In addition, questionnaires are also issued on Facebook groups, which are related to online live-stream discussions (streamers' social platform, forum for game and live-stream discussion).

The items of this research are all measured in Likert's 7-point format scale. And to assure the fitness of these questionnaire items in this research, we look back at items with good credibility and validity in prior related literatures. Moreover, introductions for items in every construct will be in section two: item measurement of research variables.

3.2   Item Measurement of Research Variables

By reviewing prior literatures related to participation intention, para-social interaction, source attractiveness model, and cognitive load, this research gives operational definitions to each research construct, and selecting or adding/removing research items which suit the scenario of online live-stream.

i. Participation Intention

This research reuses the measurement of typology of consumers' online brand-related activities (COBRA) from Muntinga et al. (2011), and modifying contexts of the items to a version suitable for online live-stream. In between these items, creating level is not included in the range of this research, thus being deleted. The third level is replaced with money participation intention from Wohn et al. (2018) (See details in the first section of Chapter 2). Furthermore, to suit the scenario of online live-stream, this research modifies the consuming level into non-creating participation, and the contributing level into contributing participation. The operative definitions are listed in Table 1.

Table 1. The operational definition and for participation intention

Construct	Operational definition	Reference
Participation intention	Live-stream viewers express their supports to streamers through various ways	Muntinga et al. (2011) Wohn et al. (2018)

ii. Para-Social interaction

In prior literatures, Experience of Parasocial Interaction Scale (EPSI-Scale) (6 items) from Hartmann and Goldhoorn (2011) was a commonly used scale for measurement of para-social interaction. ESPI-Scale measures viewers' imaginations of para-social interactive experiences between themselves and media characters. It focuses on the interactive experiences perceived by viewers. This research decides to make use of EPSI-Scale proposed by Hartmann and Goldhoorn (2011), and sorting out operational definitions as in Table 2.

Table 2. The operational definition for para-social interactive experience

Construct	Operational definition	Reference
Para-Social interaction	Influenced by streamers, the level of immediate interactions with streamers perceived one-sided by live-stream viewers	Rubin and McHugh (1987) Hartmann and Goldhoorn (2011)

iii. Source Credibility Model

Ohanian (1990) considered that source credibility was constructed with three variables: expertise, trustworthiness, and physical attractiveness. And source credibility scale, proposed by them, was a basic scale for prior advertisement marketing and propagation research. Besides Ohanian's (1990) scale, item contexts from other literatures of source credibility are also referred (e.g. Ahearne et al. 1999; Shen et al. 2010). Operational definitions are also modified to suit the scenario of online live-stream. Result of the arrangement is shown as Tables 3, 4 and 5.

Table 3. The operational definition for expertise

Construct	Operational definition	Reference
Expertise	The level of live-stream viewers' thoughts on whether streamers are familiar with discussed contents or not	Ohanian (1990) Shen et al. (2010)

Table 4 The operational definition for trustworthiness

Construct	Operational definition	Reference
Trustworthiness	Live-stream viewers' perceptions on how much trustworthy streamers are	Ohanian (1990) Ahearne et al. (1999)

Table 5 The operational definition for physical attractiveness

Construct	Operational definition	Reference
Physical attractiveness	Live-stream viewers' thoughts on how much charming or appreciated streamers' physical appearances are	Ohanian (1990) Ahearne et al. (1999)

iv. Cognitive Load

Cognitive load refers to a level of human's cognitive system loading due to limited working memories, when they are during learning process or on specific jobs (Paas and VanMerriënboer 1994). In Lee's (2017) research on user experience of mobile interface applications, the aforementioned scale was modified into the scenario of mobile application interface, and its items were reduced down to 4 from the original 6 items, through performing exploratory factor analysis. And the construct was named cognitive load. This research reuses items from Lee (2017), modifying item contexts to suit the scenario of online live-stream, and then arranging them with an operational definition into Table 6.

Table 6 The operational definition for cognitive load

Construct	Operational definition	Reference
Cognitive load	The level of cognitive system loadings perceived by viewers due to interface design of live-stream platforms	Cao et al. (2009) Paas and Van Merriënboer (1994)

3.3   Data Analyzing Method

For the retrieved data, this research first will use statistic software SPSS 24.0 to perform a descriptive statistic analysis to understand sample structures of the entirety like genders, ages, use experiences of live-stream platforms, etc. Outliers, not fully answered or too much consistently answered samples will be deleted. Then, using statistic software SmartPLS 3.2.7 to test reliability and validity of scales from measurement model, with further analysis on structural model.

For the prospect of reliability, Cronbach's α and composite reliability are both bigger than recommended standard 0.7, which means items consists of high consistency (Fornell and Larcker 1981). And for the prospect of validity, factor loadings of each item are all bigger than 0.8 and average variance extracted (AVE) for each construct are all bigger than 0.5, which means there is enough convergent validity (Fornell and Larcker 1981). And square roots of AVE for each construct are all bigger than other constructs' correlation coefficients, which means good discriminant validity (Chin 1998). Finally, structural model will be verified through partial least squares (PLS) from structural equation modeling (SEM).

4   Excepted Results

Live-stream belongs to the range of rather new industries. This research uses para-social interaction and source credibility model to discuss this rather new field. Not only it can expend literatures related to para-social interaction and source credibility model being used on live-stream industries, it also discusses the influence from live-stream platforms themselves towards relationships between streamers and viewers. Understanding influences between live-stream platforms, streamers and viewers is the expected aca-demical contribution of this research.

And through these research results, businesses of live-stream platforms can understand that design of their platforms is an attention-needed detail to viewers. Useful platforms will make viewers immerse more into live-streams. Not only they can make viewers willing to contribute to streamers, but streamers' willingness of using these platforms will also increases. Moreover, for streamers, they can understand what factors from their live-streams can influence viewers' immersions. Higher immersions can possibly increase viewers' participation intentions. When viewers are willing to make practical participations or money-wise participations, streamers can gain benefits because of this. Streamers can also increase their popularities to attract more viewers. And all of above is the expected practical contribution of this research.

Appendix

Items for measurement		
Non-creating participation	NP1	I may pay attention to every move from the streamer during live-streams
	NP2	I may check out individual information of the streamer on the live-stream platform
	NP3	I may carefully browse the streamer's conversations with other viewers
	NP4	I may support the streamer through pressing like buttons (e.g. heart, emojis, follow)
Contributing participation	CP1	I may interact with the streamer through commenting
	CP2	I may share links to live-streams of the streamer on my social platform profiles
	CP3	I may share links to live-streams of the streamer to my friends to express my support
	CP4	I may share everything I see about the streamer on chats towards him/her (e.g. his/her billboard advertisements, online videos of his/her endorsement, news about him/her)
Money participation	MP1	If the streamer decides to raise funds, I will donate money to him/her
	MP2	I may donate to the streamer to support his/her continuing streaming in the future
	MP3	I may donate to the streamer to approve his/her previous hard works
	MP4	I may give the streamer virtual gifts to express my appreciation towards him/her

Items for measurement	
When I am watching the streamer's live-stream, I feel that…	
PS1	He/she notices me
PS2	He/she knows me being here
PS3	He/she knows that I notice him/her
PS4	He/she knows that I am following him/her
PS5	He/she knows that I respond to him/her directly
PS6	He/she makes responds to me directly

Items for measurement	
TW1	I consider the streamer always treat me honestly
TW2	I consider the streamer a person I can trust
TW3	I consider the streamer always treat me straight
TW4	I think that the streamer never tries to mislead me

Items for measurement	
PA1	I think that the streamer is good-looking
PA2	I think that physical appearance of the streamer is charming
PA3	I think the streamer may be considered physically attractive
PA4	I think the streamer may be commonly considered beautiful/handsome/elegant/strong

Items for measurement	
When I am watching the streamer live streaming and using some functions on live-stream platform (e.g. commenting, pressing heart button, sharing, donating, and sending gifts).	
CL1	It does not take me extra hard work
CL2	It is not really a trouble for me
CL3	I do not feel nervous because of this
CL4	I do not think it is a hard thing to do

Items for measurement	
ET1	I think the contents that the streamer talks about is very informative
ET2	I think the contents that the streamer talks about is very professional
ET3	I think the streamer is very experienced with the contents that he/she talks about
ET4	Compared with other streamers, I think the streamer knows a lot about the contents that he/she talks about

References

Lee, T.: User Experience of Mobile Application's Interface: Measurement Development. Department of Information Management, National University of Kaohsiung (2017)

Ahearne, M., Gruen, T.W., Jarvis, C.B.: If looks could sell: moderation and mediation of the attractiveness effect on salesperson performance. Int. J. Res. Mark. **16**, 269–284 (1999)

Amos, C., Holmes, G., Strutton, D.: Exploring the relationship between celebrity endorser effects and advertising effectiveness: a quantitative synthesis of effect size. Int. J. Adv. **27**(2), 209–234 (2008)

Cao, A., Chintamani, K.K., Pandya, A.K., Ellis, R.D.: NASA TLX: software for assessing subjective mental workload. Behav. Res. Methods **41**(1), 113–117 (2009)

Chen, C., Lin, Y.: What Drives Live-Stream Usage Intention? The Perspectives of Flow, Entertainment, Social Interaction, and Endorsement. Telemat. Inform. **35**(1) 293–303 (2017)

Chin, W.W.: The Partial Least Squares Approach for Structural Equation Modeling. Modern methods for business research, pp. 295–336. Lawrence Erlbaum Associates Publishers, Mahwah (1998)

Dibble, J.L., Hartmann, T., Rosaen, S.F.: Parasocial interaction and parasocial relationship: conceptual clarification and a critical assessment of measures. Hum. Commun. Res. **42**(1), 21–44 (2016)

Erdogan, B.Z.: Celebrity endorsement: a literature review. J. Mark. Manag. **15**(4), 291–314 (1999)

Fornell, C., Larcker, D.F.: Structural equation models with unobservable variables and measurement error: algebra and statistics. J. Mark. Res. **18**(3), 382–388 (1981)

Gong, W., Li, X.: Engaging fans on microblog: the synthetic influence of parasocial interaction and source characteristics on celebrity endorsement. Psychol. Mark. **34**(7), 720–732 (2017)

Hartmann, T., Goldhoorn, C.: Horton and Wohl revisited: exploring viewers' experience of parasocial interaction. J. Commun. **61**(6), 1104–1121 (2011)

Hilvert-Bruce, Z., Neill, J.T., Sjöblom, M., Hamari, J.: Social motivations of live-streaming viewer engagement on twitch. Comput. Hum. Behav. **84**, 58–67 (2018)

Horton, D., Wohl, R.R.: Mass communication and para-social interaction. Psychiatry **19**(3), 215–229 (1956)

Hovland, C.I., Janis, I.L., Kelley, H.H.: Communication and Persuasion; Psychological Studies of Opinion Change. Yale University Press, New Haven (1953)

Hu, M., Zhang, M., Wang, Y.: Why do audiences choose to keep watching on live video streaming platforms? an explanation of dual identification framework. Comput. Hum. Behav. **75**, 594–606 (2017)

Lahtinen, L.: Mobile Information Visualisation: Recommendations for Creating Better Information Visualisation Interfaces on Mobile Devices. Retrieved from, School of Computer Science and Communication (CSC), KTH (2017). http://kth.divaportal.org/smash/get/diva2:1118715/FULLTEXT01.pdf

Madden, T.J., Ellen, P.S., Ajzen, I.: A comparison of the theory of planned behavior and the theory of reasoned action. Pers. Soc. Psychol. Bull. **18**(1), 3–9 (1992)

Merkle, E.R., Richardson, R.A.: Digital dating and virtual relating: conceptualizing computer mediated romantic relationships. Fam. Relat. **49**(2), 187–192 (2000)

Muntinga, D.G., Moorman, M., Smit, E.G.: Introducing COBRAs: exploring motivations for brand-related social media use. Int. J. Adv. **30**(1), 13–46 (2011)

Ohanian, R.: Construction and validation of a scale to measure celebrity endorsers' perceived expertise, trustworthiness, and attractiveness. J. Adv. **19**(3), 39–52 (1990)

Paas, F.G.W.C., Van Merriënboer, J.J.G.: Instructional control of cognitive load in the training of complex cognitive tasks. Educ. Psychol. Rev. **6**(4), 351–371 (1994)

Pan, P.L., Zeng, L.: Parasocial interactions with basketball athletes of color in online mediated sports. Howard J. Commun. **29**(2), 192–211 (2018)

Rubin, R.B., McHugh, M.P.: Development of parasocial interaction relationships. J. Broadcast. Electron. Media **31**(3), 279–292 (1987)

Shen, Y.-C., Huang, C.-Y., Chu, C.-H., Liao, H.-C.: Virtual community loyalty: an interpersonal-interaction perspective. Int. J. Electron. Commerce **15**(1), 49–74 (2010)

Siekkinen, M., Kämäräinen, T., Favario, L., Masala, E.: Can you see what i see? quality-of-experience measurements of mobile live video broadcasting. ACM Trans. Multimed. Comput. Commun. Appl. Article **14**, 1–23 (2018)

Singh, S., Sidhu, J.: An approach for determining trustworthiness of individuals in a web-based social network. Arab. J. Sci. Eng. **41**(2), 461–477 (2015)

Smith, T., Obrist, M., Wright, P.: Live-streaming changes the (video) game. In: Proceedings of the 11th European Conference on Interactive TV and Video - EuroITV 2013, 131 (2013)

Wohn, D.Y., Freeman, G., McLaughlin, C.: Explaining viewers' emotional, instrumental, and financial support provision for live streamers. In: Proceedings of the 2018 CHI, pp. 1–13 (2018)

Xiang, L., Zheng, X., Lee, M.K.O., Zhao, D.: Exploring consumers' impulse buying behavior on social commerce platform: the role of parasocial interaction. Int. J. Inf. Manage. **36**(3), 333–347 (2016)

Post-purchase Dissonance of Mobile Games Consumer

Shu-Chen Yang, Rui-Min Chang, and Chia-Jung Hsu[(⊠)]

Department of Information Management, National University of Kaohsiung,
Kaohsiung, Taiwan
henryyang@nuk.edu.tw, cm491648@gmail.com,
m1073311@mail.nuk.edu.tw

Abstract. Mobile game is the largest gaming platform nowadays. Most of mobile games adopt Freemium model selling virtual product in games as their revenue model. While research shows Freemium mobile games have an average life span of only ninety days, many players left the games soon after entering the game. One of factors that players left games is post-purchase dissonance. Post-purchase dissonance refers to a state that consumers feel regretful, frustrated or think they have made a wrong decision after the purchase. Prior research found that post-purchase dissonance will negatively influence satisfaction or lead to spread negative word-of-mouth.

In this study, we use expectation confirmation theory to examine which purchase motivation would cause greater dissonance on mobile game consumer. 18 in-game purchase motivations come from prior research and in-depth discussions with industry experts. The questionnaire is used to investigate on mobile games consumer in Taiwan.

Two-step cluster analysis was conducted and 18 in-game purchase motivations were divided into four cluster according to the level of post-purchase dissonance: Social & functional motivation, Affective value motivation, Impulsive buying motivation, and Speculative motivation. It is found that Speculative motivation and Impulsive buying motivation would cause greater post-purchase dissonance. Result of the research can help us understanding the antecedents of post-purchase dissonance on mobile games consumer and serve as a reference for game developers to design game mechanism.

Keywords: Mobile games · Post-purchase dissonance · Virtual product · Free to play · Purchase motivation · Expectation confirmation theory

1 Research Background

1.1 Research Background

Mobile games can be separated into two types: premium or free-to-play. A premium game refers to players must pay for the game before they can play it. A free-to-play mobile game's financial incomes mainly rely on advertisement clicking and in-app purchase (IAP). With such games, players can download them for free, playing most contents in the games, with additional contents and functions being unlocked through

J. C.-W. Lin et al. (Eds.): MISNC 2019, CCIS 1131, pp. 93–105, 2019.
https://doi.org/10.1007/978-981-15-1758-7_8

in-app purchase. This business model is known as freemium (Hanner and Zarnekow 2015). According to a report from The Wall Street Journal (2015)[1], freemium is already a common business model for apps, indicating the freemium type of mobile games has already become the trend of mobile game markets (Brockmann, Stieglitz and Cvetkovic 2015). For now, competitions in mobile game markets is heated up, making how to attract players to keep playing and purchasing the biggest challenge for game industries (Hamari et al. 2017).

1.2   Research Motivation and Purpose

Since 2003, charging methods for online games have slowly turned from subscriptions based on time period into freemium models which mainly sell virtual products. Such phenomenon attracted many researchers to have researches on gaming behaviors of players in virtual worlds (e.g., Greengard 2011).

After an accumulation for almost 20 years, there are already many researches on gaming/willingness to use (e.g., Hamari and Keronen 2017). There are also many researchers discussing factors influencing in-game purchases (e.g., Hamari et al. 2017). However, most of the prior researches about purchasing virtual products in-game discuss reasons for players willing to purchase and repurchase from a positive angle. Example like Lee et al. (2015) made a research on how gender, age, and type of game can influence on players' purchase willingness and repurchase willingness. Few of them discussed what makes players stop purchasing and leave the games they play (e.g., Merikivi et al. 2017).

There are many reasons which make players leave their games. Some of them can be due to actual feelings after payments do not match their expectations. This is caused by so-called post-purchase dissonance. Post-purchase dissonance refers to after consumers made their purchase decisions, they wonder whether if they made a right choice, at the same time, in a state of having unpleasant emotions like regrets and frustrations (Festinger 1957; Sweeney et al. 2000). Based on the view points of expectation confirmation theory from Oliver (1980), the bigger disconfirmation between consumers' expectation towards products and perceived performance, the more likely to trigger negative emotions like dissonance.

Hamari et al. sorted out nineteen in-game purchase motivations for players through triangulation, including avoiding spam, continuing play, and personalization, etc. To game developers, knowing which motivation cannot fulfill expectations thus causing post-purchase dissonances can be helpful in running their game business, and avoiding players who already made in-game purchases leaving the games due to disappointments, further reaching the goal of maintaining the rates of active players.

As mentioned above, this research hopes to be based on the researches made by Hamari et al., while referring to other prior literatures related to in-game purchase motivation. It is based on post-purchase dissonance described by expectation confirmation theory, and to find out which gap of purchase motivation is between expectation and actual feeling, further influencing consumers' post-purchase dissonances. Finally,

---

[1] https://www.wsj.com/articles/apple-google-and-amazon-reach-freemium-agreement-1422630548.

it will propose some practical suggestions based on the result of research, and serve as a reference for game developers to design purchase mechanism.

2 Literature Review

2.1 In-game Purchase Motivation and Post-purchase Dissonance

Post-purchase dissonance came from Festinger's cognitive dissonance theory (1957). Cognitive dissonance refers to a state of a person having two or more conflict or non-compatible cognitions simultaneously. At the said condition, he/she may be under psychological uncomfortableness (Festinger 1957). And post-purchase dissonance refers to when a consumer made his/her purchase decisions, doubting himself/herself whether he/she made right decisions or not, and at the same time, under a state of psychological displeasure like regret, or depression (Sweeney et al. 2000).

Dissonance includes cognitive elements and emotional elements (Festinger 1957). Another concept like dissonance is "Satisfaction". Satisfaction is defined as an emotional reaction after judging differences between product performance and corresponding standard (Westbrook and Oliver 1991). The difference between dissonance and satisfaction is that dissonance comes after decisions, while satisfaction being a final judgement after an entire process of decision making. In addition, the consequence of dissonance is unknown, which can cause consumers' continuous depressions and affect the generating of satisfaction (Sweeney et al. 2000; Koller and Salzberger 2007).

Many factors may cause the level of post-purchase dissonance. According to Oliver's expectation confirmation theory (1980), the bigger disconfirmation between a consumer's expectation before purchasing the product and his/her perceived performance after usage of the product, the bigger chance of dissonance happening, which may further influencing following satisfactions (Oliver 1993).

There was not a reliable way to directly measure consumers' dissonances in the past, only indirect measuring existed. Until a multi-aspects scale being developed by Sweeney et al. (2000), consumers' dissonances can now be measured with structural tools. The scale divides post-purchase dissonance into three aspects: emotional, wisdom of purchase, and concern over deal. The following empirical researches (e.g., Koller and Salzberger 2007) also prove this scale having great reliability and validity, making itself available for following researches on post-purchase dissonance. Therefore, this research will also use the scale developed by Sweeney et al. (2000) as a tool to measure consumers' dissonances.

2.2 In-game Purchase Motivation and Post-purchase Dissonance

Rezaei and Ghodsi (2014) used consuming value theory (Sweeney and Soutar 2001) to discuss their discovery of how emotional value, economical value, and quality value can influence paying willingness. In addition, Rauschnabel et al. (2017) made a research about Pokemon Go, finding out social norms and nostalgia can also influence players' purchase intentions.

This research will sort out prior game consumption researches, which are about virtual merchandise purchase, and in-game purchase motivation. It will be based on the 19 in-game purchase motivations from Hamari et al. (2017), with the addition of 4 purchase motivations, which were not included in Hamari's research: visual attraction (Guo and Barnes 2011; Merikivi et al. 2017), nostalgic emotion (Rauschnabel et al. 2017), social influence (Guo and Barnes 2011; Rauschnabel et al. 2017), and status/image (Jin et al. 2017). the motivation of becoming the best is also modified into performance enhancement. There are 23 purchase motivations in total.

Players' motivations of buying in-game merchandises are consumers' expectations. They expect gaining certain values through their purchases (Wang and Chang 2014). According to expectation confirmation theory, after consumers make their purchases, they will evaluate products' performances. If products' performances do not match consumers' expectations, post-purchase dissonances may occur (Oliver 1980). However, not every purchase motivation can cause post-purchase dissonance. For instance, the indulging children from the list of purchase motivations refers to a purchase which parents make to award their children or to pass the time. In this situation, parents are simply payers, they do not get involved with actual in-game experiences. So naturally, gaps between expectations and perceptions do not exist. Therefore, this research views the gap between expectations and perceptions from expectation confirmation theory as the cause for post-purchase dissonance. With references of purchase motivation from Hamari et al. (2017) and other literatures, it discusses what gaps between expectations from purchase motivations and perceived performances for consumers in-game can possibly cause their post-purchase dissonances. On a theoretical aspect, the result of this research may help understanding the reasons of post-purchase dissonance's occurrence in scenarios of virtual merchandises purchases and in-game purchases; On a practical aspect, it can help game developers as a reference when they design game purchase mechanisms, and manipulate players' purchase behaviors.

3   Research Methodology

This research's process can be divided into two phrases by its contents: questionnaire development and questionnaire issue. The first phrase is mainly developing items of the final questionnaire, assuring research scenarios based on research motivation and purpose. After collecting literatures related to in-game purchase motivation and interviewing with experts, the contents of the questionnaire are sorted out, and a pilot test of small scale has been performed to assure consistency of sentences. After that was the second phrase: issuing questionnaires, through a form of online questionnaire on platforms like Facebook, and PTT. The retrieving of questionnaires consisted of deleting invalid samples first, which was to avoid sabotaging the result by excluding non-suitable samples. After that was performing a two-step cluster analysis and encoding the data. The purchase intentions of mobile games would later be classified by their levels of post-purchase dissonance and named. Finally, theoretical and practical meanings would be given depending on the analyzed result.

3.1   Questionnaire Design

This phrase was mainly about developing contents of the questionnaire. We first referred current literatures about in-game purchases and virtual product purchases. This allowed us to sort out a list of collected in-game purchase intentions for mobile games. Based on the research from Hamari et al. (2017), two senior employees of mobile game industries and three players are invited to an in-depth interview and checking the list of purchase motivations, in order to avoid lack of literature discussions. The interview was performed through a half-structured method, the purpose was to find out if any purchase motivation was not included in literatures. Some motivations were deleted due to similar or duplicated contexts. These were all to perfectly make a list of in-game purchase motivations.

According to the result of literature discussions and the interview, purchase motivations of this research had been sorted out: enhancing performance (Alha et al. 2014; Hamari et al. 2017), continuous playing (Hamari et al. 2017; Jin et al. 2017), avoiding repetitiveness (Hamari et al. 2017), accelerating process (Hamari and Lehdonvirta 2010), protecting achievement (Hamari et al. 2017; Hamari and Lehdonvirta 2010), adding socialness (Lin et al. 2017), personalization (Wang and Chang 2014), participating special event (Hamari et al. 2017; Lehdonvirta 2009), showing off (Lehdonvirta 2009), reasonable price (Rezaei and Ghodsi 2014), supporting good game (Alha et al. 2014; Hamari et al. 2017), special pack (Saleh 2012), investing on gaming hobby (Hamari et al. 2017), unlocking content (Hamari and Lehdonvirta 2010), relieving stress, trying luck, collaborated content, and feedback. We then developed these contents into the questionnaire. The questionnaire had been tested in a small scale before the final issue. The pilot test used the convenience sampling method, and its subjects were 15 master students of the department of Information Management, National University of Kaohsiung. The purpose was to figure out if contexts of the questionnaire had literal unclearness or sentences being not smooth enough. Finally, to avoid respondents misunderstanding the questionnaire's context and causing errors in the result, sentences had been modified based on feedbacks of the pilot test.

3.2   Data Collection

In this phrase, questionnaires would be issued, and data would be analyzed. The population of this research was set as consumers of mobile game in Taiwan. The approach of online questionnaire was used in this research. The researchers made a website for the questionnaire by themselves, and then posting the link to the questionnaire in two steps, with one step being posting the link on Facebook, the other one being posting link on the biggest BBS forum in Taiwan: PTT. In addition, to give respondents more incentive, this research provided vouchers for convenient stores as prizes of a lucky draw and gave PTT dollars to every respondent from PTT as a reward.

The questionnaire of this research was divided into three parts. The first part was purchase motivation of mobile game consumers. We asked respondents to recall previous purchase experiences, and to select from the 18 purchase motivations if they ever made their purchase because of certain motivations. The second part was the measurement of post-purchase dissonance. To avoid respondents' cognitive loads due to too many questions to answer, the system would randomly select "up to three" motivations which respondents selected from the first part. Then they would be asked about the level of their post-purchase dissonances. The items of the second part came from the scale of consumers' post-purchase dissonances, developed by Sweeney et al. (2000) and Sweeney and Soutar (2006). its contents had three aspects: the emotional aspects with 6 items, wisdom of purchase with 4 items, and concern over deal with 3 items. That were 13 items in total. All of them were measured in Likert seven point scale, ranging from score 1 (strongly disagree) to score 7 (strongly agree).

This research used SPSS v2.2 to perform data analysis. After retrieving questionnaires and deleting invalid samples, the average score of each construct of respondents' different purchase dissonance motivations was calculated. After deleting samples which average score of constructs was less than 4, only samples with higher post-purchase dissonances were kept. Finally, motivations were classified and named through a two-step cluster analysis.

Cluster algorithms can be roughly divided into hierarchical clustering and non-hierarchical clustering. this research used clustering to perform classification, and using ward's method as the formula for estimation of distance between groups. The most common non-hierarchical clustering method is K-Means clustering. Before using such method, the number of groups must be determined first, which is also k-value. K-Means clustering will first randomly decide central points of clustering, in an amount of k. Then, the distance of each data to every central point will be calculated. The data will be allocated into the closest cluster, and new central points will be recalculated. These steps will be repeated until central points change no more.

Since we didn't know characteristics of the purchase motivations which caused high post-purchase dissonance, to avoid researchers' subjective thoughts influencing the result of clustering, we thus used the approach of two-step cluster analysis. That was to perform hierarchical clustering first. And after deciding suitable amount of clustering, we then used K-Means clustering to classify 18 motivations, and discussing characteristic of each cluster later.

4 Data Analysis

4.1 Descriptive Statistics

The date of issuing final questionnaires of this research was at Apr 11th, 2018 to Apr 30th, 2018. 776 questionnaires have been retrieved. After excluding 48 not fully answered ones, 728 valid questionnaires were left. Within the retrieved samples, the proportion of gender was 55:45; Over 70% of the respondents' ages were between 19 to 29; Their highest education levels were commonly bachelor/faculty degrees (69.2%); Occupations wise, full-time job employees and students were the most, 44.5% and 39.7% respectively. For the aspect of mobile game playing as a hobby, near 70% of the respondents played mobile games every day; They played for over 1 h per day, while over 2 h (30.6%) and over 30 min but not over 1 h (25.4%) took the most part of them. Nearly half of respondents' disposable Incomes were below NT$10,000; purchases on mobile games wise, most respondents' average money spent on mobile games per month was below NT$500, that was 61.5% of them. Overall, about 85.3% respondents' expenses on mobile games were not over NT$1,000.

In addition, purchase motivations wise, the top 3 selected ones were: reasonable price (572), special pack (538), and enhancing performance (534). Over 70% of the respondents had these three purchase motivations; the 3 least selected purchase motivations were: adding socialness (98), protecting achievement (180), and showing off (184). Among these three, adding socialness only took about 10% of the respondents. That made it the least chosen purchase motivation.

4.2   Cluster Analysis

During the process of answering questionnaires of this research, respondents were asked to recall their past purchase experiences, and ticking boxes of "I used to make in-game purchases because of that." from the given 18 motivations. Due to respondents having different purchase experiences, with the addition of random selection from the system, Differences happened on the number of samples for each motivation.

Since this research would like to discuss the characteristic of motivations which caused high post-purchase dissonance on consumers, within averages of the constructs, we deleted samples (with average score smaller than 4) which relatively did not appear to have post-purchase dissonance happened to them.

After some exclusions, purchase intelligence still held the highest average score among the three constructs; It was, at the same time, the construct with the most samples being kept. This indicated that a relatively large number of samples' average scores were higher in the construct of purchase intelligence. And the number of kept samples in emotional reaction and deal suspicion were very close to each other.

4.2.1   Hierarchical Cluster Analysis

Due to the process of deleting samples with low dissonance, the amounts kept in each motivation and each construct varied. To calculate influence from the number of kept samples with high dissonance, this research multiplied the average score of each motivation construct by the amount of its samples. That gave us a total score for each motivation construct. Based on the result, a hierarchical cluster analysis was performed. The result of hierarchical cluster analysis would stop merging if suddenly, the value of agglomerative coefficient increased massively. When clusters merged from 4 to 3, the coefficient increased massively, so the merging was stopped. The best amount of clusters was 4.

4.2.2   K-Means Cluster Analysis

After given the best number of clusters, we classified 18 purchase motivations into 4 groups based on each motivation's constructs like emotional reaction, purchase intelligence, and deal suspicion.

This research named each cluster based on the characteristics of motivations within the cluster. The names and naming reasons are shown as below:

- Cluster 1, with 6 motivations: adding socialness, showing off, investing on gaming hobby, feedback, protecting achievement, and avoiding repetitiveness. The characteristics of these motivations in this cluster vary, and most of them are not directly related to the contents of games. Therefore, this cluster was named social & functional motivation.
- Cluster 2, with 6 motivations: relieving stress, personalization, accelerating process, collaborated content, supporting good game, and unlocking content. The motivation in this cluster are mostly to fulfill certain emotional needs of consumers. This kind of motivation reflects consumers' internal prospective for value. Some motivations are also directly related to consumers' emotions (relieving stress). Therefore, this cluster was named affective value motivation.
- Cluster 3, with 5 motivations: participating special event, continuous playing, reasonable price, special pack, and enhancing performance. This cluster can easily make players purchase impulsively; and it is common to see enhancing performance and continuous playing in mobile games. These are motivations which stimulate players to purchase. Therefore, this cluster was named impulsive buying motivation.
- Cluster 4, with only 1 motivation: trying luck. Trying luck is, among 18 motivations, the one with the highest construct average scores and the highest number of samples with high dissonance. It shows obvious differences with other motivations. Trying luck includes gambling elements like chance and uncertainty. Therefore, this cluster was named speculative motivation.

5 Discussion

5.1 Analysis Result and Discussion

This research used a survey technique. From the aspect of in-game purchase motivation, it discussed the phenomenon of players' post-purchase dissonances on freemium mobile games. The result includes common purchase motivations in games, the levels of dissonance caused by each motivation, and the characteristics of motivations which can easily cause high post-purchase dissonance. Below is the discussion based on the data analysis result, with the proposition of theoretical and practical meanings, research restrictions, and future research recommendations.

5.1.1 Consumers' Purchase Motivations in Mobile Games and Post-purchase Dissonance

According to statistic result, it can be discovered that among the 18 motivations sorted by this research through literatures and an interview, purchase count for each motivation differs a lot. the top 3 purchase counts belong to these motivations: "reasonable price", "enhancing performance", and "special pack". All of them are common in game to stimulate players to make their purchases. However, in this research, adding socialness is the least selected one among all motivations (with only 92 out of 728 respondents' choses, approximately 13%). This research supposed that, all modern mobile games can provide basic functions to fulfill players' social needs. Therefore, less players make their purchases because of this.

For the aspect of post-purchase dissonance, among the retrieved samples, overall post-purchase dissonance of mobile game is not obvious, with three constructs' average scores being all smaller than 4. However, there are still some consumers having high post-purchase dissonances. The result and prior literatures indicate: post-purchase dissonance may happen at any phrase during purchasing process. But not every consumer has corresponding post-purchase dissonance phenomenon after their purchases (Oliver 1993; Sweeney et al. 2000).

5.1.2 Purchase Motivation's Characteristic Making High Dissonance

According to prior research, negative effects on merchandise satisfaction (Oliver 1980; Sweeney et al. 2000) drop on consumers with high post-purchase dissonances. Therefore, for the developers, to keep their games running for a long period, it needs to be controlled to not making high post-purchase dissonances to consumers. This research focuses on players' purchasing motivations triggered by games' contents, and discussing which purchasing motivations can possibly cause higher levels of consumers' post-purchase dissonances.

According to the result of two-step cluster analysis, we classified 18 in-game purchase motivations into 4 groups, based on the levels of post-purchase dissonance they caused and their characteristics. These groups are: 1. social & functional motivation, 2. affective value motivation, 3. impulsive buying motivation, and 4. speculative motivation. These 4 clusters, apart from speculative motivation, the other three's motivations scores on purchase intelligence are all obviously higher than scores on other constructs of post-purchase dissonance like emotional reaction and deal suspicion. Furthermore, according to the research from Saleh (2012), consumers with hedonic personalities make less consideration on whether they need this merchandise or not when they are making purchase decisions (that is: purchase intelligence). And after their purchases, they may suffer from high levels of regret. Therefore, this research considers that, within entertaining informative systems like mobile games, players are commonly prone to purchase impulsively due to games' stimulations and being in pursuits of pleasure, and thus having higher levels of post-purchase dissonance.

Here are discussions focusing on four clusters respectively:

Social & functional motivation: this kind of motivation can make the original purchase goals be reached immediately after purchases being made. According to expectation confirmation theory from Oliver (1980), the gap between expectation and perception is smaller, which means less conflicts on cognition. Therefore, the level of post-purchase dissonances made by motivations from social & functional motivation cluster is lower.

Affective value motivation: based on their emotions, consumers are willing to spend money on games they think are good (supporting good game), or characters, items which they consider worthy (collaborated content, personalization). However, if these contents can not match consumers' inner expectations, post-purchase dissonances may happen.

Impulsive buying motivation: the main purpose of players playing games is pursuing pleasure. To maintaining their joyfulness, players usually spend money on their games (enhancing performance and unlocking content). Besides, developers often create anxiety for players and make them purchase through promotion strategies. These factors easily cause players' impulsive purchase behaviors, with easily happened regret, making post-purchase dissonances like doubting decisions or considering these merchandises unnecessary.

Speculative motivation: Currently, many mobile games start to implement mechanisms like lucky draw. The post-purchase dissonances caused by such mechanisms are different from ones caused by other motivations, along with stronger negative emotional effects. This research considers that consumers have high expectations and hopes before their lucky draws, thinking themselves able to win rare prizes, but only a huge gap with disappointments is made after prizes being revealed. Therefore, compared with other kinds of motivations, speculative motivation is the one can most easily cause post-purchase dissonance, and with the most amount of samples in this research.

5.2   Theoretical and Practical Meaning

5.2.1   Theoretical Meaning

The types of merchandises for post-purchase dissonance ranges from past offline physical merchandises (e.g. Engel 1963) to recent physical merchandises in the environment of electric commercial (e.g., Koller and Salzberger 2007). But virtual merchandises like in-game items were never discussed. This research saw freemium mobile games, which mainly sold virtual merchandises, as a target for discussion. And it was found out that even virtual merchandises did not have physical appearances, and having different natures with service-type merchandises, the phenomenon of post-purchase dissonance still existed. It is still worth to discuss players' or virtual product consumers' post purchase dissonances.

In addition, when in-game purchase and virtual merchandise purchase were discussed by prior literatures, few of them made a research on players' specific purchase motivations caused by gaming designs (e.g., Wang and Chang 2013). This research is based on the research from Hamari et al. (2017), with the addition of an interview with professionals in game industries and veteran players. 18 in-game purchase intentions have been sorted out, and they can help researchers to understand the making and stimulations of in-game purchases from the personal aspect of consumers.

According to result of the hierarchical cluster analysis, it can be found out that among the 18 purchase motivations, in-game speculative motivation and impulsive buying motivations can all easily make players have negative emotions after their purchases, which means higher levels of players' post-purchase dissonances. Affective value motivations fulfill players' needs of contents they like, which causes fewer negative emotions, thus lower levels of post-purchase dissonance. The in-game social & functional motivations can mostly become effective as soon as purchases being completed, thus overall causing the lowest levels of dissonance.

5.2.2 Practical Meaning

Post-purchase dissonance can have negative influence on satisfaction and repurchase intention. However, when it comes to mobile games, the motivations which can easily cause high post-purchase dissonance (e.g. trying luck) can possibly lead to more purchases due to players' mindsets of sunk costs. The result cannot be all that negative. whether game developers can manipulate levels of players' post-purchase dissonances to trigger more purchases or not is very worth being discussed.

Prior researches considered that to lower the chance of consumers' post-purchase dissonances happen, increasing their product knowledges (Guo et al. 2016), reducing anxieties and uncertainties during purchases (Menasco and Hawkins 1978), actively communicating with consumers after purchases (Milliman and Decker 1973), and offering powerful after-sale warranties (Bawa and Kansal 2008) can all be done practically. From the aspect of gaming design, speculative motivation can stimulate players to make massive in-game purchases, but it also is the easiest approach to cause players' post-purchase dissonances. To reduce the post-purchase dissonances caused by such motivation, there are a few ways to do. First, through indicating number or chances to win for each prize, it can make players avoid feeling fooled due to sunk cost misunderstandings and non-clearly shown chances. Second, giving players with bad lucks some compensations, for instance: mechanisms of guarantee and transformation. These two mechanisms can comfort some players after a chain of failed draws, this reducing their negative emotions or regrets.

According to the result of this research, even for those who suffer from high post-purchase dissonances, the levels of post-purchase dissonance caused by social and functional motivations and emotional motivations are still lower. Therefore, developers should focus on how to build good gaming community ecosystems, increasing players' social values and making a game IP (intellectual property) which can fulfill players' every emotional need. In addition, Cohen and Goldberg (1970) thought that when consumers trusted brands more, post-purchase dissonance would be less likely generated. Therefore, developers may also build some feedback mechanisms for communications between players and them, through their official communities, and this increasing players' approvals and trusts towards their brands.

5.3 Limitations and Future Research

There are four main restrictions for this research. First is the time restriction: this research investigated respondents' post-purchase dissonances on one single time period. However, prior researches (e.g., Koller and Salzberger 2012) indicated that, as time went by, levels of consumers' post-purchase dissonances would also be decreased. And within the questionnaire of this research, respondents were asked to recall previous purchase experiences, thus making it unable to accurately measure the states of dissonances at those time periods. It is recommended to perform a two-phrase surveying test for future researches. This will allow researchers to compare consumers' current levels of post-purchase dissonance to the levels surveyed after one period of time.

The second recommendation is the issue of purchase motivation classification. The in-game purchase motivations this research discusses are inspired and based on gaming design. However, no connection is shown between each purchase motivation. How to extract constructs of players' purchasing motivations and classify them more effectively is what future researches can discuss.

The third recommendation is influences from types of games and players' experiences: prior researches (e.g., Hanner and Zarnekow 2015) indicated that on different genres of games, consumers' purchase motivations can also be strong or weak, and their purchase behaviors are not necessarily the same. Therefore, differences may happen on levels of post-purchase dissonance, too. This research did not make discussion specifically for any genres of games, thus influences from some players who are prone to purchase impulsively towards post-purchase dissonances on specific game genres is not discovered. It is recommended for future related researches that to make consideration of differences on different game genres. In addition, Guo et al. (2016) considered that with more years of players' gameplaying, it brings richer game-related knowledges, and thus not able to easily cause post-purchase dissonances. Therefore, it is recommended to survey respondents' experiences on gaming for future researches.

The fourth recommendation is the problems with samples' representations. This research only targeted mobile game consumers of Taiwan region to perform an investigation. However, since cultures differ in every country, players' preferences to game genres may differ, too. The result may not be suitable for consumers of every age and from every country. Also, over 70% of respondents in this research are not over 30 years old, which makes the result unable to explain samples of middle age appropriately.

References

Alha, K., Koskinen, E., Paavilainen, J., Hamari, J.: Free-to-play games: Professionals' perspectives. In: Proceedings of Nordic Digra 2014, Gotland, Sweden, pp. 1–14 (2014)

Bawa, A., Kansal, P.: Cognitive dissonance and the marketing of services: some issues. J. Serv. Res. 8(2), (2008)

Brockmann, T., Stieglitz, S., Cvetkovic, A.: Prevalent business models for the Apple app store. In: Wirtschaftsinformatik, pp. 1206–1221 (2015)

Cohen, J.B., Goldberg, M.E.: The dissonance model in post-decision product evaluation. J. Mark. Res. 7(3), 315–321 (1970)

Engel, J.F.: Are automobile purchasers dissonant consumers? J. Mark. 27(2), 55–58 (1963)

Festinger, L.: A theory of cognitive dissonance. Stanford University, Stanford, CA (1957)

Greengard, S.: Social games, virtual goods. Commun. ACM 54(4), 19–22 (2011)

Guo, J., Huang, E., Lin, S.C.: Investigating the Effect of Pre-Purchase Search and Ongoing Search on Post-Purchase Dissonance (2016)

Guo, Y., Barnes, S.: Purchase behavior in virtual worlds: an empirical investigation in Second Life. Inf. Manag. 48(7), 303–312 (2011)

Hamari, J., Alha, K., Järvelä, S., Kivikangas, J.M., Koivisto, J., Paavilainen, J.: Why do players buy in-game content? An empirical study on concrete purchase motivations. Comput. Hum. Behav. 68, 538–546 (2017). https://doi.org/10.1016/j.chb.2016.11.045

Hamari, J., Lehdonvirta, V.: Game design as marketing: how game mechanics create demand for virtual goods. Int. J. Bus. Sci. Appl. Manag. 5(1), 14–29 (2010)

Hanner, N., Zarnekow, R.: Purchasing behavior in free to play games: Concepts and empirical validation. In: 2015 48th Hawaii International Conference on System Sciences, pp. 3326–3335. IEEE (2015)

Jin, W., Sun, Y., Wang, N., Zhang, X.: Why users purchase virtual products in MMORPG? An integrative perspective of social presence and user engagement. Internet Res. **27**(2), 408–427 (2017). https://doi.org/10.1108/IntR-04-2016-0091

Koller, M., Salzberger, T.: Cognitive dissonance as a relevant construct throughout the decision-making and consumption process-an empirical investigation related to a package tour. J. Customer Behav. **6**(3), 217–227 (2007)

Koller, M., Salzberger, T.: Heterogeneous development of cognitive dissonance over time and its effect on satisfaction and loyalty. J. Customer Behav. **11**(3), 261–280 (2012)

Lee, J., Lee, J., Lee, H., Lee, J.: An exploratory study of factors influencing repurchase behaviors toward game items: a field study. Comput. Hum. Behav. **53**, 13–23 (2015). https://doi.org/10.1016/j.chb.2015.06.017

Lehdonvirta, V.: Virtual item sales as a revenue model: identifying attributes that drive purchase decisions. Electr. Commer. Res. **9**(1–2), 97–113 (2009). https://doi.org/10.1007/s10660-009-9028-2

Lin, X., Featherman, M., Sarker, S.: Understanding factors affecting users' social networking site continuance: A gender difference perspective. Inf. Manag. **54**(3), 383–395 (2017)

Menasco, M.B., Hawkins, D.I.: A field test of the relationship between cognitive dissonance and state anxiety. J. Mark. Res. **15**(4), 650–655 (1978)

Merikivi, J., Tuunainen, V., Nguyen, D.: What makes continued mobile gaming enjoyable? Comput. Hum. Behav. **68**, 411–421 (2017)

Milliman, R.E., Decker, P.J.: The use of post-purchase communication to reduce dissonance and improve direct marketing effectiveness. J. Bus. Commun. **27**(2), 159–170 (1973)

Oliver, R.L.: A cognitive model of the antecedents and consequences of satisfaction decisions. J. Mark. Res. **17**(4), 460 (1980). https://doi.org/10.2307/3150499

Oliver, R.L.: Cognitive, affective, and attribute bases of the satisfaction response. J. Consum. Res. **20**(3), 418 (1993). https://doi.org/10.1086/209358

Rauschnabel, P.A., Rossmann, A., tom Dieck, M.C.: An adoption framework for mobile augmented reality games: The case of Pokémon Go. Comput. Hum. Behav. **76**, 276–286 (2017)

Rezaei, S., Ghodsi, S.S.: Does value matters in playing online game? An empirical study among massively multiplayer online role-playing games (MMORPGs). Comput. Hum. Behav. **35**, 252–266 (2014). https://doi.org/10.1016/j.chb.2014.03.002

Saleh, M.A.E.H.: An investigation of the relationship between unplanned buying and post-purchase regret. Int. J. Mark. Stud. **4**(4), 106–120 (2012). https://doi.org/10.5539/ijms.v4n4p106

Sweeney, J.C., Soutar, G.N.: A short form of Sweeney, Hausknecht and Soutar's cognitive dissonance scale (2006)

Sweeney, J.C., Hausknecht, D., Soutar, G.N.: Cognitive dissonance after purchase: a multidimensional scale. Psychol. Mark. **17**(5), 369 (2000). https://doi.org/10.1002/(SICI)1520-6793(200005)17:5%3c369:AID-MAR1%3e3.3.CO;2-7

Sweeney, J.C., Soutar, G.N.: Consumer perceived value: the development of a multiple item scale. J. Retail. **77**(2), 203–220 (2001). https://doi.org/10.1016/S0022-4359(01)00041-0

Wang, W.T., Chang, W.H.: The integration of the expectancy disconfirmation and symbolic consumption theories: a case of virtual product consumption. In: 46th Hawaii International Conference on System Sciences, pp. 2949−2956. IEEE

Wang, W.T., Chang, W.H.: A Study of virtual product consumption from the expectancy disconfirmation and symbolic consumption perspectives. Inf. Sys. Frontiers **16**(5), 887–908 (2014). https://doi.org/10.1007/s10796-012-9389-2

Westbrook, R.A., Oliver, R.L.: The dimensionality of consumption emotion patterns and consumer satisfaction. J. Consum. Res. **18**(1), 84–91 (1991)

Japanese University Students' Acceptance of Cross-border Electronic Commerce

Takashi Okamoto[1]([✉]) [iD], Jiro Yatsuhashi[2] [iD], and Naoki Mizutani[3] [iD]

[1] Ehime University, Matsuyama, Ehime, Japan
okamoto.takashi.me@ehime-u.ac.jp
[2] Kanagawa University, Yokohama, Kanagawa, Japan
jiro@kanagawa-u.ac.jp
[3] Okayama University of Science, Okayama, Japan
mizutani@mgt.ous.ac.jp

Abstract. With increasing smartphone penetration, online purchasing has become a common and representative purchasing channel. The development of the Internet, enabling online transactions, has led to the emergence of cross-border electronic commerce. Despite the fact that online shopping sites are developing very quickly, the factors that influence consumers' purchase intention (PI) on cross-border online shopping sites are not clear. Furthermore, prior studies have not identified the perceived differences between domestic and cross-border online shopping sites.

This study examines young people's acceptance of both domestic and cross-border online shopping sites, and identifies the relationships among the variables involved in online shopping sites. This study indicates that the PI with online shopping sites is directly influenced by perceived usefulness (PU), trust (TR), and credibility of personal information protection (PIP) on online shopping sites. In addition, this study reveals the perceived differences between domestic and cross-border online shopping sites. In particular, credibility of PIP and TR of cross-border online shopping sites strongly influences PI, and PU weakly influences PI. On the other hand, PU, PIP, and TR of domestic online shopping sites have a similar influence on PI.

Keywords: Cross-border electronic commerce · Purchase intention · Trust of online market · Credibility of personal information protection · Structural equation model

1 Introduction

With increasing smartphone penetration, online purchasing has become a common and representative purchasing channel. The overall scale of Japanese business-to-consumer (B to C) electronic commerce reached 16 trillion yen in 2017 [1]. The B to C electronic commerce market has become large, and popular among most Japanese people.

The development of the Internet, enabling online transactions, has led to the emergence of a new model of foreign trade: cross-border electronic commerce [2]. Cross-border electronic commerce has risen and developed into an important model in the rapidly expanding electronic commerce market [3].

© Springer Nature Singapore Pte Ltd. 2019
J. C.-W. Lin et al. (Eds.): MISNC 2019, CCIS 1131, pp. 106–117, 2019.
https://doi.org/10.1007/978-981-15-1758-7_9

Despite the fact that online shopping sites are developing very quickly, the factors that influence consumers' purchase intention (PI) on cross-border online shopping sites are not clear. Little research has explored the individual consumer's psychological processes with respect to signing up, and the benefits derived from purchasing on cross-border shopping sites [4]. Furthermore, prior studies have not identified the perceived differences between domestic and cross-border online shopping sites.

Although many prior studies have examined and revealed the issues of transaction intention or PI in B to C models, only a few studies have attempted to explain the factors influencing the adoption of cross-border online shopping sites. Successful marketing strategies are based on understanding consumers, their characteristics, the benefits they pursue in the products they buy, and the channels or platforms they choose from [4].

This study has two main objectives. The first objective is to indicate the prevalence of both domestic and cross-border online shopping among young people. The second objective is to reveal the perceived differences between domestic and cross-border online shopping sites. To fulfill these research objectives, this study administered questionnaires. We examined the respondents' perceptions of both domestic and cross-border online shopping sites using structural equation modeling (SEM). The model that this study examines, is based on the extended technology acceptance model (TAM) to adapt to domestic and cross-border online shopping sites [5].

Our research demonstrates that about 94.7% of respondents have purchased goods via domestic online shopping sites, that is, Japanese online shopping sites. On the other hand, 60.6% of respondents had never shopped online on cross-border online shopping sites. Cross-border online shopping have not become a popular purchasing channel, although domestic online shopping has already become a representative purchasing channel for young people.

This study reveals the perceived differences between domestic and cross-border online shopping sites. In particular, the credibility of PIP and TR in cross-border online shopping sites strongly influences PI, and PU weakly influences PI. On the other hand, PU, PIP, and TR of domestic online shopping sites have a similar influence on PI. These findings imply that improvement of TR and PIP of cross-border online shopping sites for Japanese people will promote Japanese utilization of the cross-border online shopping sites.

2 Factors Affecting Online Shopping Sites

Many studies have profiled consumers' PI and examined consumers' perceptions of and behavior regarding online shopping. Many electronic commerce adoption studies have used an extended TAM to examine consumers' PI. The model that this study examines is based the model to examine purchasing intention in flea markets [5, 6]. This prior model is also based on extended TAM.

Extended TAM has two major determinants, PU and perceived ease of use (PEOU), which influence behavioral intention [7]. Prior studies indicate that PEOU and PU significantly influence individual attitudes toward online shopping [8, 9]. PU is positively associated with continuance intention toward mobile commerce in most electronic commerce studies [10, 11]. PEOU is defined as the degree to which a person believes that using a particular system would be free of effort [7]. Many studies suggest that PEOU has a positive effect on PU.

Some prior studies include perceived TR and perceived SQ as factors to explain consumer acceptance of online shopping [12, 13]. SQ, or the factors related to the nature of technologies, services, or systems, were added in other studies [14].

Prior studies have reported that consumer attitudes toward online shopping are determined by TR, and hence, TR has been used by many electronic commerce researchers [13, 15]. TR plays an important role in all business relationships, especially cross-border electronic commerce. The separation of both time and space in cross-border electronic commerce brings information asymmetry to buyers, which increased risks and uncertainty. This highlights the importance of TR. Increases in the level of TR directly and positively affect PI of online shopping [16–18]. Many studies have determined that TR has a positive and significant impact on electronic commerce. Several studies have found that there is a positive relationship between user TR, and PU and PEOU [13, 15, 16, 18].

Many online shoppers utilize their smartphones, and store their personal information on them. Since smartphones have become essential devices for daily life, users need secure services, especially the credibility of PIP. Security or privacy issues have been considered as important factors in using mobile services and Internet-based commerce [8, 19, 20]. The perceived security risks, which lead to insufficient protection of their personal information, negatively affect PI and TR.

3   Model and Proposed Hypotheses

Electronic commerce has become common for almost all consumers. They can utilize cross-border online shopping sites, as these sites have similar user interfaces and payment means as domestic online shopping sites. Technological barriers for cross-border e-commerce have been getting lower. Most Japanese people, however, do not utilize cross-border online shopping sites, and have internal barrier against such sites. Differences of perceptual structure may exist between using Japanese and cross-border online shopping sites.

The current research is based on TAM' thus, the PU of shopping on online shopping sites, and PEOU of online shopping sites are included. Therefore, the first two hypotheses of this study are as follows:

H1-1:   The PU of online shopping sites significantly and positively influences the PI of consumers.

H2-1:   The PEOU of online shopping sites significantly and positively influences the PU of the online shopping sites.

This study extends TAM by adding three more constructs that may affect its primary variables and PI. TR and PIP are added, and their impact on PI in online shopping sites is studied.

Technology-related factors become significant, affecting PU, PEOU, and other extended constructs [6, 19, 21]. This study calls these factors "SQ".

Consumers' perception of unexpected result of products and services is referred to as perceived risk. Consumers' hesitation toward online shopping can be biased by these risk factors [12]. Perceived risk has a negative effect on consumers' PI on online shopping sites. Therefore, user TR in online shopping sites may affect PEOU, PU, and PI. The credibility of PIP may also affect PI and TR in the online shopping sites.

This study's four hypotheses regarding SQ are as follows:

H3-1:   The SQ of online shopping sites significantly and positively influences the PU of online shopping sites.

H3-2:   The SQ of online shopping sites significantly and positively influences the PEOU of online shopping sites.

H3-3:   The SQ of online shopping sites significantly and positively influences the TR in online shopping sites.

H3-4:   The SQ of online shopping sites significantly and positively influences the credibility of PIP provided by the online shopping sites.

This study's three hypotheses regarding TR in online shopping sites are as follows:

H4-1:   TR of online shopping sites significantly and positively influences the PU of online shopping sites.

H4-2:   TR of online shopping sites significantly and positively influences the PEOU of consumers.

H4-3:   TR of online shopping sites significantly and positively influences the PI of consumers.

This study's two hypotheses regarding the credibility of PIP provided by online shopping sites are as follows:

H5-1:   The PIP provided by online shopping sites significantly and positively influences the PI of consumers.

H5-2:   The PIP provided by online shopping sites significantly and positively influences the TR on online shopping sites.

4   Materials and Method

4.1   Collection of Data

The target population of this research was young people in Japan, who have shopped online or will do so in the future. Our respondents comprised university students. The research was conducted in January 2018 and used a self-administered questionnaire. Our valid respondents were 322 students registered at Ehime University, Matsuyama University, and Kanagawa University. Table 1 reports respondents' demographic details. As all students were majoring in the social sciences, we were able to limit the number of demographic factors, other than the university to which they belonged.

There were minor deviations in gender and grade. All respondents were familiar with smartphones and the Internet.

Table 1. Respondent composition

Demographic details	Categories	Number of respondents
Gender	Female	151 (46.9%)
	Male	171 (53.1%)
Grade	1st	134 (41.6%)
	2nd	88 (27.3%)
	3rd	72 (22.4%)
	4th and over	28 (8.7%)
University	Ehime University	79 (24.5%)
	Matsuyama University	44 (13.7%)
	Kanagawa University	199 (61.8%)

4.2 Measures and Data Analysis Method

Our questionnaire assessed behavior and perceptions toward online shopping sites or malls. Questions on behavior were about online purchasing experiences on domestic online shopping sites, and cross-border online shopping sites. The measurements of their value structures consisted of 19 questions. The items assessing the various constructs are listed in Table 2. Most items in the questionnaire were obtained from questionnaires used in prior studies [5]. All constructs were measured using a five-point Likert-type scale starting from (1) "disagree" to (5) "agree". Each respondent answered these questions for both domestic online shopping sites and cross-border online shopping sites. Demographic data pertaining to gender and grade were sought on a nominal scale.

To test the hypothesized relationships in the operation model, SPSS 22 was used to check the reliability of the model constructs. Our research utilized SPSS Amos 25 software to conduct the structural equation modeling analyses.

5   Results

5.1 Demographics of the Respondents

Table 3 shows the respondents' experiences with respect to online purchasing on both, Japanese and cross-border online shopping sites. Regarding shopping experiences on domestic online shopping sites, 95% of respondents had shopped online on domestic online shopping sites. About 43% percent of respondents had purchased goods via online shops several times in the past year, and 30% of them purchase online once a

Table 2. List of construct indicators

Constructs	
PU	
Purchasing with online shopping sites is convenient	PU1
Purchasing with online shopping sites is easy	PU2
Purchasing with online shopping sites is useful	PU3
PEOU	
Using online shopping sites or malls is easy for me	PEOU1
It is easy for me to become skilled at using online shopping sites	PEOU2
Online shopping sites have a concise and easy-to-understand layout	PEOU3
SQ	
Online shopping sites have a substantial information providing function for	SQ1
ordering and delivery	SQ2
Online shopping sites have a substantial search function	SQ3
I can get information about various goods	SQ4
Online shopping sites have substantial online reviews	
TR	
Information about goods on online shopping sites are accurate and reliable	TR1
Goods on online shopping sites are reliable and not fake	TR2
I do not have anxiety about shopping on online shopping sites	TR3
PIP	
Online shopping sites protect my privacy	PIP1
Online shopping sites have a reliable PIP system	PIP2
Online shopping sites protect my payment information	PIP3
PI	
I would like to purchase goods on online shopping sites	PI1
I would like to purchase goods more often on online shopping sites	PI2
I would recommend purchasing goods on online shopping sites to others	PI3

month. These results imply that domestic online shopping sites has become a representative purchasing channel for young Japanese people.

Regarding shopping experiences on cross-border online shopping sites, 61% of respondents had never shopped online on cross-border online shopping sites. Only 27% of them had purchased goods via cross-border online shopping sites several times in a year. These results imply that very few respondents have shopped at cross-border online shopping sites, compared to shopping experiences on domestic online shopping sites. Cross-border shopping behavior has not become common for Japanese young people.

Table 3. Respondents' experience with respect to online purchasing on domestic or cross-border shopping sites

	Categories	Number of respondents
Average number of purchases on domestic online shopping sites	Never used	17 (5.3%)
	Several times in a year	140 (43.5%)
	Once in a month	98 (30.4%)
	Several times in a month	55 (17.1%)
	Once in a week	9 (2.8%)
	Twice in a week	2 (0.6%)
	Three times or over in a week	1 (0.3%)
Average number of purchases on cross-border shopping sites	Never used	195 (60.6%)
	Several times in a year	88 (27.3%)
	Once in a month	27 (8.4%)
	Several times in a month	7 (2.2%)
	Once in a week	4 (1.2%)
	Twice in a week	0 (0.0%)
	Three times or over in a week	1 (0.3%)

Table 4. Summary of reliability analysis

Construct	Cronbach's alpha: for domestic online shopping sites	Cronbach's alpha: for cross-border online shopping sites
PU	0.706	0.907
PEOU	0.776	0.891
SQ	0.711	0.904
TR	0.797	0.897
PIP	0.919	0.959
PI	0.842	0.925

5.2 Reliability of Measurement and Model Fit

In order to measure the internal consistency of the scales, reliability of measurement was tested using Cronbach's alpha [16, 17]. Table 4 provides the Cronbach's alpha values for all constructs; all values were greater than 0.70 for both Japanese and cross-border online shopping sites, and were thus, deemed acceptable. The results indicate that all constructs are reliable and have high internal consistency.

Our research model was validated using SEM. The results of the model fit the indices, indicating that the data fits the model. Regarding online purchasing on domestic online shopping sites, absolute fit measures of x^2/df were found to be 2.445. The model has a goodness-of-fit Index (GFI) of 0.902, normed fit index (NFI) of 0.894, comparative fit index (CFI) of 0.934, and root mean square error of approximation (RMSEA) of 0.067. These measures revealed a reasonable model fit.

Regarding online purchasing at cross-border online shopping sites, absolute fit measures of x^2/df were found to be 3.423. The model has GFI of 0.868, NFI of 0.923, CFI of 0.944, and RMSEA of 0.087. Although GFI was under 0.9, these measures also revealed a reasonable model fit.

5.3   Hypothesis Testing Regarding Online Purchasing on Domestic Online Shopping Sites

A path diagram for a structural model is shown in Fig. 1. A one-way path indicates a significant relationship. The estimated coefficients, and the results of the hypotheses tests are provided in Table 5. All coefficients are significant at the 1% level.

Table 5. Test of hypotheses on domestic online shopping sites

Hypotheses		Structural coefficient	t-value	Result
H1-1	PU->PI	0.326	5.325	***
H2-1	PEOU->PU	0.484	5.417	***
H3-1	SQ->PU	0.510	4.487	***
H3-2	SQ->PEOU	0.501	4.675	***
H3-3	SQ->TR	0.310	3.995	***
H3-4	SQ->PIP	0.477	5.529	***
H4-1	TR->PU	−0.205	−2.645	***
H4-2	TR-PEOU	0.221	2.821	***
H4-3	TR->PI	0.273	3.268	***
H5-1	PIP->PI	0.324	4.224	***
H5-2	PIP->TR	0.540	7.860	***

***: Significant at the 1% level

Table 5 indicates that all hypotheses are significant at the 1% level for online purchasing on domestic online shopping sites. For H4-1, the results indicated that TR has a significant and negative effect on PU. All other hypotheses had a positive effect. In particular, SQ showed weak positive effects on TR as compared to PEOU, PU, and PIP. Thus, SQ is not an important determinant of TR. For H4, TR had a negative effect on PU and thus, TR is a negative determinant for PU. These results confirm that TR directly works to reduce PU, whereas TR increases PU indirectly via PEOU. For H5, PIP has a stronger positive effect on TR than SQ, and thus, PIP is an important determinant of TR. PI, PU, TR, and PIP have similar positive effects on PI. The results confirm that PU, TR and PIP have a similar impact on PI, and are important determinants of PI.

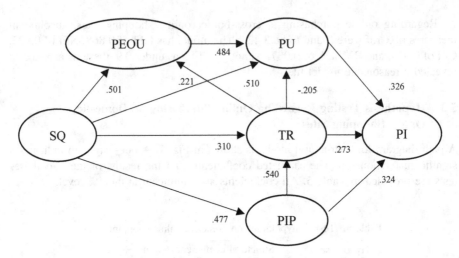

Fig. 1. Results of SEM for online purchasing on domestic online shopping sites

5.4   Hypotheses Testing Regarding Online Purchasing on Cross-border Online Shopping Sites

Regarding online purchasing on cross-border online shopping sites, a path diagram for a structural model is shown in Fig. 2. The estimated coefficients and results of the hypotheses test are provided in Table 6. All coefficients are significant at the 1% level.

Table 6 indicates that all hypotheses are significant for online purchasing on cross-border online shopping sites. For H4-1, the results indicated that TR has a significant and negative effect on PU, while the other hypotheses have a positive effect similar to domestic online shopping sites. Most results of cross-border online shopping sites are similar to the results of domestic online shopping sites. PU, TR, and PIP, however, have different effects on PI from that in the case of domestic online shopping sites. TR and PIP have strong effects on PI, whereas PU has weak effects on PI. This result confirm that TR and PIP are important determinants of PI. PU is not an important determinant of PI, as compared to TR and PIP. SQ has a strong effect on PEOU, as compared to domestic online shopping sites. These results are significantly different from the result of domestic online shopping sites.

5.5   Perceptional Differences Between Domestic and Cross-border Online Shopping Sites

Regarding domestic online shopping sites, SQ has strong effects on PEOU, PU, and PIP. In the case of cross-border online shopping sites, SQ has a stronger effect on PEOU than on PU or PIP. This result implies that SQ is the important determinant of PEOU in the case of cross-border online shopping sites. Thus, the improvement of SQ positively and strongly influences PEOU, and indirectly influences PU and PI. Most Japanese people may find it difficult to use cross-border online shopping sites, even if they are accustomed to using domestic online shopping sites. They may face the problems of different language, payment, user interface, shipping, and so on. Perceived system qualities will influence the PEOU of online shopping sites.

For cross-border online shopping sites, PU has a weak effect on PI as compared to domestic online shopping sites. On the other hand, TR and PIP have strong effects on PI. Japanese people may think that trust and personal information protection are most important factors while purchasing from cross-border online shopping sites. Most Japanese people may overestimate the risks of shopping on cross-border online shopping sites, because they are not familiar with such sites.

These results are significantly different from the results of domestic online shopping sites.

Table 6. Test of hypotheses on cross-border online shopping sites

Hypothesis		Structural coefficient	t-value	Result
H1-1	PU->PI	0.160	3.559	***
H2-1	PEOU->PU	0.393	4.498	***
H3-1	SQ->PU	0.558	6.086	***
H3-2	SQ->PEOU	0.720	10.528	***
H3-3	SQ->TR	0.268	5.157	***
H3-4	SQ->PIP	0.570	10.149	***
H4-1	TR->PU	−0.201	−3.429	***
H4-2	TR->PEOU	0.152	2.787	***
H4-3	TR->PI	0.323	4.397	***
H5-1	PIP->PI	0.414	5.864	***
H5-2	PIP->TR	0.625	11.351	***

***: Significant at the 1% level

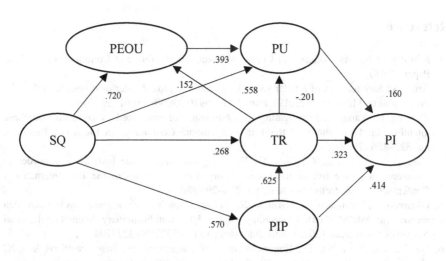

Fig. 2. Results of SEM for online purchasing on cross-border online shopping sites

6   Conclusions

This study was designed to investigate the factors influencing consumer PI in electronic commerce. TAM was used as the base theory, and extended to apply to both electronic commerce markets.

Our results support the validity of the TAM-based research model, which asserts that PI on online shopping sites is determined by PU, TR, and PIP.

This study examines young people's acceptance of both domestic and cross-border online shopping sites. It reveals the perceived differences between domestic and cross-border online shopping sites. In particular, the credibility of PIP and TR in cross-border online shopping sites strongly influences PI, while PU weakly influences PI. On the other hand, PU, PIP, and TR of domestic online shopping sites have a similar influence on PI.

This study has a limitation in that the sample is limited in its diversity, as all respondents are university students, and the sample may not be fully representative of all smartphone users. In future studies, the respondents should be more diverse, so as to generalize the research results. In this study, behavioral intention was measured using the extended TAM methodology. Thus, the model in this study has somewhat general structure. It can be applied to not only cross-border online shopping but also other technology acceptance issues. Further studies are needed to investigate the influence of other external variables of extended TAM for both domestic and cross-border electronic commerce. The familiarity of Internet usage, experience of shopping on cross-border online shopping sites, familiarity of cross-border language, and income can be considered as variables.

Acknowledgements. This research was supported by the Ministry of Education, Science, Sports and Culture, Grant-in-Aid for Scientific Research (C), 18K01798, 2018–2021.

References

1. Ministry of Internal Affairs and Communication: Information and Communication White Paper (2018)
2. Artur, S.: Key features of e-tailer shops in adaption to cross-border e-commerce in the EU. Sustainability **11**(6), 1589 (2019). https://doi.org/10.3390/su11061589
3. Zhiyong, L., Yang, L.: Investigating on determinants of cross-border e-commerce purchase intention. In: Proceedings of the 18th International Conference on Electronic Business, pp. 825–826 (2018)
4. Xiaoyu, X., Fang-Kai, C., Li, E.Y.: Exploring consumer value path of cross-border e-commerce: a perspective of means-end theory. In: Proceedings of the 18th International Conference on Electronic Business, pp. 284–294 (2018)
5. Okamoto, T., Yatsuhashi, J., Mizutani, N.: Young people's purchase intentions in online flea market. In: MISNC 2018 Proceedings of the 5th Multidisciplinary International Social Networks Conference (2018). https://doi.org/10.1145/3227696.3227704
6. He, D., Lu, Y., Zhou, D.: Empirical study of consumers' purchase intentions in C2C electronic commerce. Tsinghua Sci. Technol. **13**(3), 287–292 (2008). https://doi.org/10.1016/S1007-0214(08)70046-4

7. Davis, F.D.: Perceived usefulness, perceived ease of use, and user acceptance of information technology. MIS Q. **13**(3), 319–340 (1989). https://doi.org/10.2307/249008
8. Hoque, M.R., Ali, M.A., Mahfuz, M.A.: An empirical investigation on the adoption of e-commerce in Bangladesh. Asia Pac. J. Inform. Syst. **25**(1), 1–24 (2015). https://doi.org/10.14329/apjis.2015.25.1.001
9. Suh, B., Han, I.: The impact of customer trust and perception of security control on the acceptance of electronic commerce. Int. J. Electron. Commer. **7**(3), 135–161 (2003). https://doi.org/10.1080/10864415.2003.11044270
10. Lu, J.: Are personal innovativeness and social influence critical to continue with mobile commerce? Internet Res. **24**(2), 134–159 (2014). https://doi.org/10.1108/IntR-05-2012-0100
11. Kim, D.J., Ferrin, D.L., Rao, H.R.: Trust and satisfaction, two stepping stones for successful e-commerce relationships: a longitudinal exploration. Inf. Syst. Res. **20**(2), 237–257 (2009). https://doi.org/10.1287/isre.1080.0188
12. Rind, M.M., Hyder, M., Saand, A.S., Alzabi, T., Nawaz, H., Ujan, I.: Impact Investigation of perceived cost and perceived risk in mobile commerce: analytical study of Pakistan. Int. J. Comput. Sci. Netw. Secur. **17**(11), 124–130 (2017)
13. Çelik, H.E., Yılmaz, V.: Extending the technology acceptance model for adoption of e-shopping. J. Electron. Commer. Res. **12**(2), 152–164 (2011)
14. Min, Q., Ji, S., Qu, G.: Mobile commerce user acceptance study in China: a revised UTAUT model. Tsinghua Sci. Technol. **13**(3), 257–264 (2008). https://doi.org/10.1016/S1007-0214(08)70042-7
15. Qiu, L., Li, D.: Applying TAM in B2C e-Commerce research: an extended model. Tsinghua Sci. Technol. **13**(3), 265–272 (2008). https://doi.org/10.1016/S1007-0214(08)70043-9
16. Corbitt, B.J., Thanasankit, T., Yi, H.: Trust and e-Commerce: a study of consumer perceptions. Electron. Commer. Res. Appl. **2**(3), 203–215 (2003). https://doi.org/10.1016/S1567-4223(03)00024-3
17. Gefen, D.: E-commerce: the role of familiarity and trust. Omega Int. J. Manage. Sci. **28**(6), 725–737 (2000). https://doi.org/10.1016/S0305-0483(00)00021-9
18. Kim, D.J., Ferin, D.L., Rao, H.R.: A trust-based consumer decision-making model in electronic commerce: the role of trust, perceived risk, and their antecedents. Decis. Support Syst. **44**(2), 544–564 (2008). https://doi.org/10.1016/j.dss.2007.07.001
19. Jun, J., Lee, K.J., Kim, B.G.: Factors affecting user acceptance of mobile commerce services. Asia Pac. J. Inform. Syst. **26**(4), 489–508 (2016). https://doi.org/10.14329/apjis.2016.26.4.489
20. Polla, M.L., Martinelli, F., Sgandurra, D.: A survey on security for mobile devices. IEEE Commun. Surv. Tutorials **15**(1), 446–471 (2013). https://doi.org/10.1109/SURV.2012.013012.00028
21. Matthew, K., Lee, O., Turban, E.: A trust model for consumer internet shopping. Int. J. Electron. Commer. **6**(1), 75–91 (2001). https://doi.org/10.1080/10864415.2001.11044227

Visualization of Health Data

Veronica Castro Alvarez and Ching-yu Huang[(⌂)]

School of Computer Science, 1000 Morris Avenue, Union, NJ 07083, USA
{Castroav, chuang}@kean.edu

Abstract. As data becomes more accessible, visualization methods are needed to help make sense of the information. Analyzing and visualizing data helps the public to better recognize the patterns and connections between different data-sets. By using visual elements such as graphs, charts, and maps, it is easier to see and understand the trends and outliers in data. This project aims to study the correlation between environmental factors and public health. Large sets of data pertaining to the environment and health were gathered from open data sources. The tool used to analyze and visualize the collected data is Tableau, which is a software program that is used to transform data into dashboards and visuals such as treemaps, histograms, or area charts. For this project, the data will be displayed through charts and interactive maps that will be created through this software.

Keywords: Environment · Health · Tableau · Visualization

1 Introduction

The environment can greatly impact individuals' health, and one of the biggest environmental factors that contributes to disease is pollution. There have been various studies and articles published on the effects of environmental factors on health and disease. According to an article published in The Lancet, a peer-reviewed medical journal, "Pollution is the largest environmental cause of disease and premature death in the world today. Diseases caused by pollution were responsible for an estimated 9 million premature deaths in 2015—16% of all deaths worldwide" [1]. Some of the most prominent and unresolved issues related to public health are air pollution, drinking water, soil pollution, and weather conditions [2]. Outdoor air pollution seems to be negatively affecting health outcomes at increasing rates, with respiratory diseases, such as asthma, and mortality being the most common and prevalent health outcomes [3].

These pollution factors, in addition to inadequate water sanitation, agricultural practices, built environments, and climate change all play a major role in the likelihood of developing a disease, and people are exposed to these risk factors every day in their homes and communities. There is also a need for safe water around the globe, and research has been published on the different areas of water and health, such as socioeconomy of water, water quality, water treatment, water microbiology, water sanitation, and water resources [4]. For example, study conducted in 26 Sub-Saharan African countries found that the "time spent walking to a household's main water source was found to be a significant determinant of under-five child health" [5].

© Springer Nature Singapore Pte Ltd. 2019
J. C.-W. Lin et al. (Eds.): MISNC 2019, CCIS 1131, pp. 118–130, 2019.
https://doi.org/10.1007/978-981-15-1758-7_10

Another important aspect to the burden of disease and death is income level, as individuals with low economic standing and those in developing countries usually have poor water quality and access [6]. "Nearly 92% of pollution-related deaths occur in low-income and middle-income countries and, in countries at every income level, disease caused by pollution is most prevalent among minorities and the marginalised" [1]. Additionally, individuals living at a low-income level are more at risk of death from disease in countries where they do not have any access to or have limited access to healthcare. For example, a recent study on the gradient between child's health and family income in Canada did not find that children from low-income families suffered more from poor health than children from high-income families (contrary to previous studies from the United States). However, "the contrast between Canadian and U.S. children may reflect the effects of universal health insurance in Canada" [7].

When comparing low-income and middle-income countries, both power equality and per capita income seem to be important factors in relation to health performance, "while only power equality appears to be a factor in explaining population health in middle-income countries" [8]. However, neither power equality nor per capita income seem to be a factor in high-income countries, but rather "expenditures on health services relative to GDP and sanitation access" appear to be factors in these countries [8].

Although there have been research studies published on the areas of public health and the environment, there have not been many studies which use visualization methods to display the data associated with both of these areas. Big data visualization can be used to communicate large amounts of information that otherwise may not be as easy to understand by a general audience without a more in-depth look. This study aims to expand upon previous research of this topic by visualizing the available open data in order to make the information easier to understand for a general public audience.

2   Methods

Data was retrieved from two online open data sources, the World Health Organization's Global Health Observatory and World Bank Open Data. The datasets pertaining to health factors were retrieved from the Global Health Observatory, while the datasets pertaining to environmental factors were retrieved from World Bank Open Data. The original formats of the datasets were either Comma-Separated Values (CSV) files or Microsoft Excel files.

2.1   Interpreting the Data

After gathering all of the datasets, each dataset was interpreted using Tableau's Data Interpreter, which is a feature that can be used when connecting to Excel files to clean and transform the data. Data that is provided in Excel files are usually made to be easy to read with the human interface in mind, so they may often include titles, stacked headers, or empty rows and columns to make the file more visually appealing and easy to understand. This was the case for many of the datasets used in this study. However, these visually appealing aspects of the file can make the data more difficult for Tableau to interpret. Tableau provides this data interpretation feature to help identify the structure of the data in the Excel file and detect if there are any things such as titles, stacked headers, or empty cells in the file. Once any of these aspects have been detected, Tableau's data interpreter converts the data into the proper format to be used for analysis. One of the datasets retrieved from World Bank Open Data was of Total Population by Country per Year. Figure 1 shows how this dataset appeared in Tableau before using Tableau's Data Interpreter to interpret the data, and Fig. 2 shows how the data appeared after using the Data Interpreter.

▦ ▤ Sort fields	Data source order ▾			☐ Show aliases ☐ Show hidden fields	267 → rows			
Abc	Abc	Abc	Abc	#	#	#	#	#
Data	Data	Data	Data	Data	Data	Data	Data	D
F1	**F2**	**F3**	**F4**	**F5**	**F6**	**F7**	**F8**	**F**
Data Source	World Developm...	null	null	null	null	null	null	
Last Updated Date	3/21/2019	null	null	null	null	null	null	
Country Name	Country Code	Indicator Name	Indicator Code	1,960	1,961	1,962	1,963	
Aruba	ABW	Population, total	SP.POP.TOTL	54,211	55,438	56,225	56,695	
Afghanistan	AFG	Population, total	SP.POP.TOTL	8,996,351	9,166,764	9,345,868	9,533,954	
Angola	AGO	Population, total	SP.POP.TOTL	5,643,182	5,753,024	5,866,061	5,980,417	
Albania	ALB	Population, total	SP.POP.TOTL	1,608,800	1,659,800	1,711,319	1,762,621	
Andorra	AND	Population, total	SP.POP.TOTL	13,411	14,375	15,370	16,412	
Arab World	ARB	Population, total	SP.POP.TOTL	92,490,932	95,044,497	97,682,294	100,411,076	
United Arab Emir...	ARE	Population, total	SP.POP.TOTL	92,634	101,078	112,472	125,566	
Argentina	ARG	Population, total	SP.POP.TOTL	20,619,075	20,953,077	21,287,682	21,621,840	

Fig. 1. Dataset before data interpretation

▦ ▤ Sort fields	Data source order ▾			☐ Show aliases ☐ Show hidden fields	264 → rows		
⊕	⊕	Abc	Abc	#	#	#	#
Data	Data	Data	Data	Data	Data	Data	Data
Country Name	**Country Code**	**Indicator Name**	**Indicator Code**	**1960**	**1961**	**1962**	**1963**
Aruba	ABW	Population, total	SP.POP.TOTL	54,211	55,438	56,225	5
Afghanistan	AFG	Population, total	SP.POP.TOTL	8,996,351	9,166,764	9,345,868	9,53
Angola	AGO	Population, total	SP.POP.TOTL	5,643,182	5,753,024	5,866,061	5,98
Albania	ALB	Population, total	SP.POP.TOTL	1,608,800	1,659,800	1,711,319	1,76
Andorra	AND	Population, total	SP.POP.TOTL	13,411	14,375	15,370	1
Arab World	ARB	Population, total	SP.POP.TOTL	92,490,932	95,044,497	97,682,294	100,41
United Arab Emir...	ARE	Population, total	SP.POP.TOTL	92,634	101,078	112,472	12
Argentina	ARG	Population, total	SP.POP.TOTL	20,619,075	20,953,077	21,287,682	21,62
Armenia	ARM	Population, total	SP.POP.TOTL	1,874,120	1,941,491	2,009,526	2,07
American Samoa	ASM	Population, total	SP.POP.TOTL	20,013	20,486	21,117	2
Antigua and Bar...	ATG	Population, total	SP.POP.TOTL	55,339	56,144	57,144	5

Fig. 2. Dataset after data interpretation

2.2   Pivoting the Data

After interpreting the datasets, they then had to be transposed from crosstab format (wide format) to columnar format (long format). Most of the datasets that were retrieved were originally in crosstab format, where there was one separate column for each year and each year was a column header. However, the ideal structure for a data table that is to be used for analysis is columnar format, where each row represents an observation belonging to a particular category, such as Year or Number of Cases. To transpose the tables from crosstab format to columnar format, his was done using Tableau's Pivot function, as shown in Figs. 3 and 4.

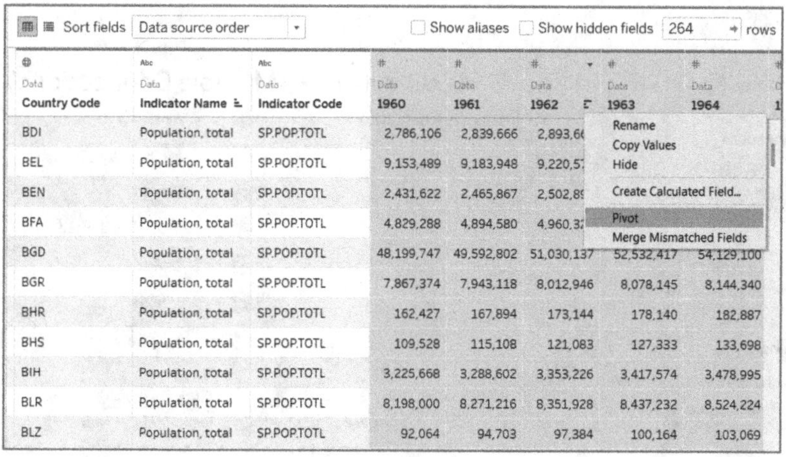

Fig. 3. Dataset before data pivot

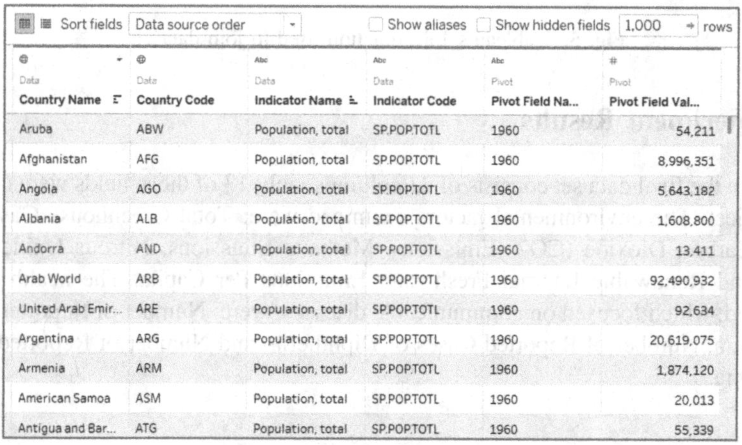

Fig. 4. Dataset after data pivot

2.3   Joining the Data

After interpreting, pivoting, and filtering the datasets, they were then joined together by the fields Country Name and Year by left join using Tableau's Join function. This function allows sperate tables that are related by specific fields, such as Country and Year, to be combined on those common fields. This joining results in a virtual table that can then be used for analyzation and visualization. Figure 5 shows how the tables containing data on Cholera, Diphtheria, Malaria, Air Pollution, Arable Land, Renewable Internal Freshwater Resources, and Crude Death Rate were joined to the table containing on Country Information.

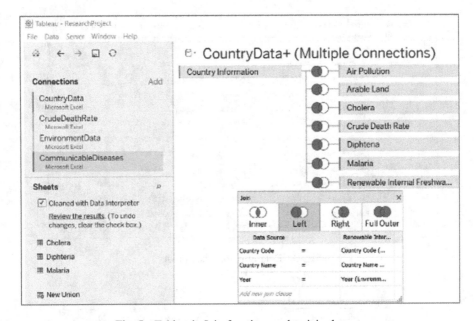

Fig. 5. Tableau's Join function used to join datasets

3   Experiment Results

Although the final data set consists of 23 columns, only 13 of those fields were used for this project. The environmental factors examined were: Total Greenhouse Gas Emissions, Carbon Dioxide (CO2) Emissions, Methane Emissions, Nitrous Oxide Emissions, and Renewable Internal Freshwater Emissions Per Capita. The health factors examined, which focused on communicable diseases, were: Number of Reported Cases of Cholera, Number of Reported Cases of Diphtheria, and Number of Reported Cases of Malaria.

The first factors that were examined for this project were the environmental factors which were gathered from World Bank Open Data [9]. Figure 6 shows the Total Greenhouse Gas Emissions (in units of kt of CO2 equivalent) by Country for the years 2000–2012. Some of the greenhouse gases that are summed together to determine this variable are Carbon Dioxide (CO2) emissions (in units of kt), Methane emissions (in units of kt of CO2 equivalent), and Nitrous Oxide emissions (in units of kt of CO2 equivalent). The following symbol map is one of the numerous charts that Tableau offers to visualize data. From the symbol map, it is shown that China emits the most greenhouse gases than any other country, followed by the United States of America.

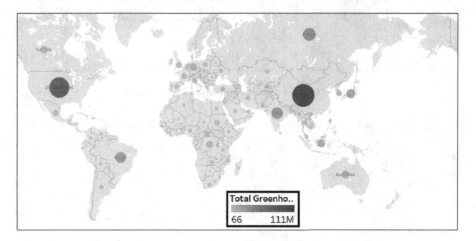

Fig. 6. Total Greenhouse Gas Emissions by Country

Tableau allows the user to zoom into specific areas of the map in order to emphasize any prominent countries or areas. Figure 7 shows the map zoomed into China, and Fig. 8 shows the map zoomed into the United States, the two countries with the highest greenhouse gas emissions. When the user hovers over the symbol, a tooltip appears to show the details for that selected country. The user can select which variables are to appear in the tooltip. In the tooltip used for this symbol map, the Country Name, CO2 Emissions, Methane Emissions, Nitrous Oxide Emissions, and Total Greenhouse Gas Emissions are shown for each country.

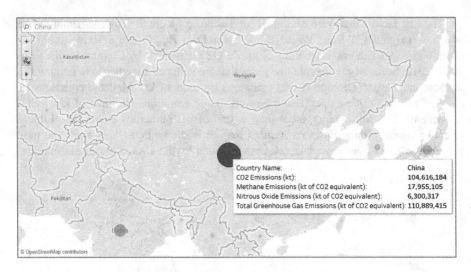

Fig. 7. China's Total Greenhouse Gas Emissions

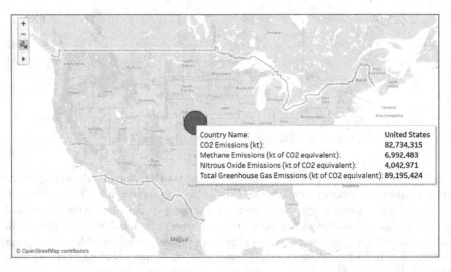

Fig. 8. United States' Total Greenhouse Gas Emissions

The second variable that was examined in this study was Population by Region. From the following bar chart, it is shown that East Asia and the Pacific has the highest population compared to any other region, followed by South Asia, and then by Sub-Saharan Africa. The region with the lowest population compared to any other region is North America, followed by the Middle East and North Africa.

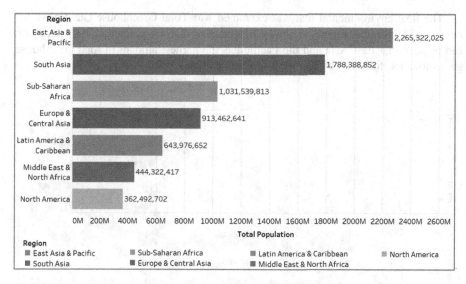

Fig. 9. Total Population by Region

The next environmental factor examined in this study was the Renewable Internal Freshwater Resources per Capita (in units of cubic meters). From the following bar chart, it is shown that the region with the highest amount of renewable internal freshwater resources per capita is Europe and Central Asia. The region with the lowest amount of renewable internal freshwater resources per capita is the Middle East and North Africa, followed by North America, and then by South Asia. Sub-Saharan Africa also has a low amount of renewable internal freshwater resources per capita.

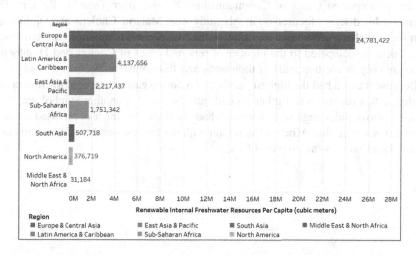

Fig. 10. Renewable Internal Freshwater Resources per Capita by Region

The final environmental that was examined was Total Greenhouse Gas Emissions over Time by Region. Figure 11 shows an area chart that was created in Tableau, and it can be seen that East Asia and the Pacific has the highest amount of total greenhouse gas emissions, followed by Europe and Central Asia.

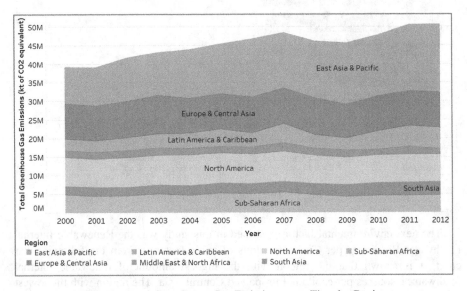

Fig. 11. Total Greenhouse Gas Emissions over Time by Region

After examining some of the environmental data, the health data gathered from the Global Health Observatory [10] was then examined. Figure 12 shows a line chart of the Number of Reported Cases of Communicable Diseases over Time by Region. The communicable diseases focused on in this study were Malaria, Cholera, and Diphtheria. From the line chart, it is shown that Malaria has the highest number of reported cases, in the millions, compared to the number of reported cases of Cholera and Diphtheria, which are only in the hundreds of thousands and thousands, respectively.

Because malaria had the highest number of reported cases compared to cholera and diphtheria, this disease was further looked into. In Fig. 13, another symbol map was created to show which regions or countries had the highest number of reported cases of malaria. It is shown that Africa is the region with the highest number of reported cases of malaria, as well as the country of India.

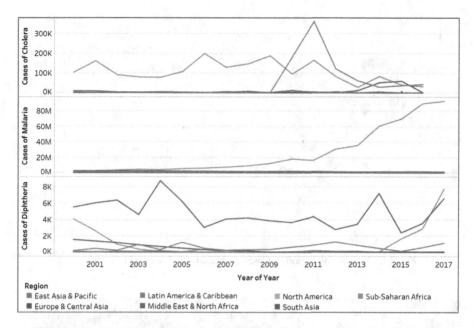

Fig. 12. Number of Reported Cases of Communicable Diseases over Time by Region

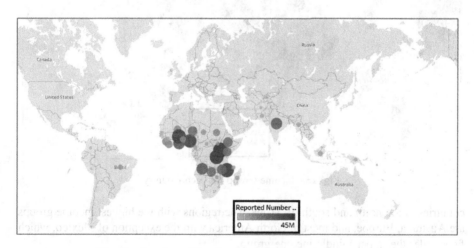

Fig. 13. Number of Reported Cases of Malaria by Country

The final factor that was examined in this study was the Income Group of each Country. Figure 15 is a map that was created in Tableau which shows which income group each country belongs to. The income groups were categorized as high income, upper middle income, lower middle income, and low income. Much of Africa falls into the low income to lower middle income group, with a few upper middle income

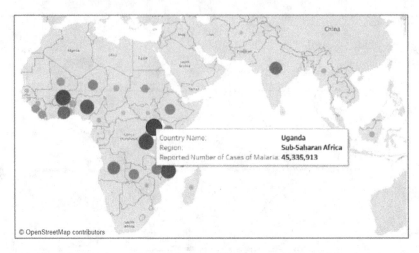

Fig. 14. Number of Reported Cases of Malaria in Africa and South Asia

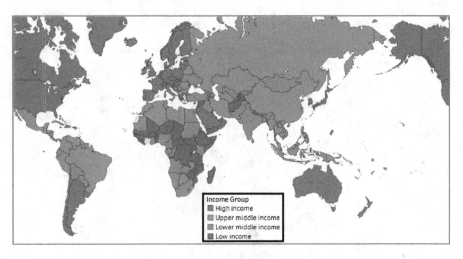

Fig. 15. Income Group for each Country

countries in the north and south of Africa. The regions with the highest income groups are Australia, Europe, and most of North America with the exception of Mexico, which falls under the upper middle income group.

4   Conclusions

The charts and maps that were created using Tableau show that the Number of Reported Cases of Communicable Diseases, specifically Malaria, may be related to both the Number of Internal Freshwater Resources per Capita, and the Income Group of each country. When comparing Fig. 10 (Renewable Internal Freshwater Resources per Capita by Country) to Fig. 13 (Number of Reported Cases of Malaria by Country), it can be concluded that the amount of renewable internal freshwater resources a country has may play a role in the number of reported cases of malaria for that country. Sub-Saharan Africa is one of the regions with the lowest amounts of renewable internal freshwater resources per capita, and is also has the highest number of reported cases of malaria. Additionally, South Asia, the region with the third lowest amount of renewable internal freshwater resources per capita, also has a high number of reported cases of malaria, especially in India.

In Fig. 10, it is shown that North America is the region with the second lowest amounts of renewable internal freshwater resources per capita. However, this region has very little to no cases of malaria, even though most of the other regions, such as Sub-Saharan Africa and South Asia, which also have low amounts of renewable internal freshwater resources per capita, have high cases of malaria. The difference between these regions may be due to the income level of the countries in those regions.

In Fig. 15 (Income Group for each Country), it is shown that most of the countries in Africa fall under the Low Income Group and Lower Middle Income Group. These countries are also the countries, as seen in Fig. 14, which have high numbers of reported cases of malaria. The countries in Africa which fall under the Upper Middle Income Group do not have the same high number of cases of malaria as the countries in the lower income groups. The cases of malaria in these upper income level countries are relatively low compared to the lower income level countries in the same continent. Additionally, North America, which has low amounts of renewable internal freshwater resources per capita, falls under the High Income Group. It appears that the high income level of this region may contribute to it not having a high number of cases of malaria as Africa does. The high income level of this region may allow the people who live in this area to be better equipped at preventing the spread of these communicable diseases, as well as treating any cases of these diseases if they should occur.

A final conclusion that can be made from this study is that air pollutions, specifically greenhouse gas emissions, do not seem to play a role in the prevalence of the three communicable diseases examined in this study. In Figs. 1 and 5, it is shown that East Asia and the Pacific, North America, and Europe and Central Asia all have high amounts of total greenhouse gas emissions, yet these regions have very little to no cases of malaria, cholera, or diphtheria. If this study were to be continued, another health factor that can be examined is non-communicable diseases, in addition to the communicable diseases that have already been examined.

Although this study showed that the prevalence of malaria, cholera, and diphtheria are highest in the regions where renewable internal freshwater resources are low, and in the countries where the income group is low, this does not necessarily imply causality between these factors. More research would have to be done in order to determine the causality between the environmental factors and the health factors used in this study.

References

1. Landrigan, P.J., et al.: The lancet commission on pollution and health. Lancet Comm. **391**, 462–512 (2017). https://doi.org/10.1016/S0140-6736(17)32345-0
2. Ban, J., et al.: Environmental health indicators for china: data resources for Chinese environmental public health tracking. Environ. Health Persp. **127**(4), 1–10 (2019). https://doi.org/10.1289/EHP4319
3. Sun, Z., Zhu, D.: Exposure to outdoor air pollution and its human health outcomes: a scoping review. PLoS ONE **14**(5), 1–18 (2019). https://doi.org/10.1371/journal.pone.0216550
4. Setty, K., et al.: Faster and safer: research priorities in water and health. Int. J. Hyg. Environ. Health **222**(4), 593–606 (2019). https://doi.org/10.1016/j.ijheh.2019.03.003
5. Pickering, A.J., Davis, J.: Freshwater availability and water fetching distance affect child health in Sub-Saharan Africa. Environ. Sci. Technol. **46**(4), 2391–2397 (2012). https://doi.org/10.1021/es203177v
6. World Health Organization: Health & environment: tools for effective decision-making (2005). https://www.who.int/heli/publications/brochure/en/
7. Wei, L., Feeny, D.: The dynamics of the gradient between child's health and family income: evidence from Canada. Soc. Sci. Med. **226**, 182–189 (2019). https://doi.org/10.1016/j.socscimed.2019.02.033
8. Torras, M.: The impact of poqer equality, income, and the environment on human health: some inter-country comparisons. Int. Rev. Appl. Econ. **20**(1), 1–20 (2006). https://doi.org/10.1080/02692170500362199
9. World Health Organization: Global Health Observatory (GHO) data (2019). https://www.who.int/gho/en/
10. The World Bank: World Bank Open Data (2019). https://data.worldbank.org/

Augmented Reality as a Reinforcement to Facilitate ESP Learning for Nursing Students

YingLing Chen[✉]

Central for General Education, Oriental Institute of Technology, Taipei,
Taiwan, R.O.C.
ci10226@mail.oit.edu.tw

Abstract. There is a great deal of research focusing on the benefit of teaching English for Specific Purpose (ESP) and yet there are relatively few studies suggesting ways how language teachers can help learners cope with learning by implementing Augmented Reality (AR). Therefore, the purpose of this study was to identify whether the integration of AR helped increasing EFL learners' learning Effectiveness and motivation. The research took place in the nursing department at a private university in the northern part of Taiwan. 48 junior nursing students participated in the study. Mixed method was applied in the study. The questionnaire of AR assists learning L2 effectiveness and satisfaction was implemented at the first section of the data collection. In addition, the use of semi-structure interview and field note techniques were applied to be the main method of gathering qualitative research data. The data was obtained through adapted and modified open-ended questions. Participants were asked to reflect on how AR operated in the ESP context and how AR created the model of interactive learning environments after each AR learning practice. The results indicated that implementing AR was a vital strategy for producing independent thinkers and learners. They flourish under ESP-driven and creative approach; nursing students will be able to develop their own inquiries.

Keywords: Augmented Reality · English as a Foreign Language · ESP

1 Introduction

Digital environments are a big part of learners' education and lifestyles. Incorporating existing technologies and techniques such as Augmented Reality as teaching and instructional support is an inevitable movement in the 21 centuries. There is a great deal of research focusing on the benefit of teaching English for Specific Purpose (ESP) and yet there are relatively few studies suggesting ways how language teachers can help learners cope with learning by implementing Augmented Reality (AR). Therefore, the purpose of this study was to identify whether the integration of AR helped increasing L2 learners' ESP learning effectiveness and motivation. The use of technology for foreign language teaching and learning especially since the late 1960s has created new opportunities and possibilities and made teachers keep up with the latest technology in order to meet the expectations regarding classroom activities in the

© Springer Nature Singapore Pte Ltd. 2019
J. C.-W. Lin et al. (Eds.): MISNC 2019, CCIS 1131, pp. 131–141, 2019.
https://doi.org/10.1007/978-981-15-1758-7_11

language learning process (Alsied and Pathan 2013; Richards 2014). In relation to the reasons about technology integration into foreign language classrooms, another researcher named Lam (2000) states that technology gives a hint about the attitude of the students and makes them more alert and enthusiastic. AR offers an innovative way to mediate learning opportunities by projecting onto the learner's physical reality. According to interactionist approaches to second language acquisition (SLA) (Hatch 1978; Long 1996), interaction is the most important way in which learners obtain data for language learning. Previous research examining different aspects of technology-assisted learning has found equivocal results discovering its effectiveness and outcomes (Hewitt-Taylor 2010). In addition, AR support can be a practical tool for educators enhance their ESP instruction and EFL learners improve their performance.

2 Literature

2.1 Augmented Reality

In the late 1960s, the use of technology for foreign language teaching and learning had created various opportunities and possibilities which made teachers keep up with the latest technology in order to meet the expectations regarding classroom activities in the language learning process (Alsied and Pathan 2013; Richards 2014). Augmented Reality (AR) visual displays, a particular subset of Virtual Reality (VR) related technologies that involve the merging of real and virtual worlds somewhere along the "virtuality continuum" which connects completely real environments to completely virtual ones (Milgram 1996, p. 2). The concept of AR had been interpreted as having both virtual spaces on the one hand and reality on the other available within the same visual display environment (Hughes et al. 2005). AR is practiced as a media to offer learners an ideal virtual space with reality essential for target curriculum. Classroom Instructions primarily focus on the acquisition of clinical knowledge and skills. Student nurses are expected to associate classroom theory into practice in order to reduce the gap between textbook material and clinical practice. Scenario-based instruction is often considered by the educators in order to minimize the gap. Among the many teaching methods and technical applications used in nursing education, clinical scenario-based simulation has become an increasing popular instruction method (McAllister et al. 2013). The interactivity of these technology environments is a very important feature for learning. Interactivity makes it easy for students to revisit specific parts of the environments to explore them more fully. (Mankin et al. 1997). Learning through real-world contexts creates the bridge for future nurses to be more experienced and knowledgeable. The concept of a "virtuality continuum" relates to the mixture of classes of objects presented in any particular display situation, as illustrated in Fig. 1, where real environments, are shown at one end of the continuum, and virtual environments, at the opposite extremum. As indicated in the figure, the most straightforward way to view an Augmented Reality environment, therefore, is one in which real world and virtual world objects are presented together within a single display (Milgram 1996, p. 3).

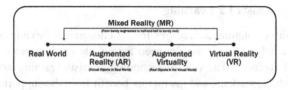

Fig. 1. Simplified representation of a "virtuality continuum". (Milgram 1996, p. 3)

2.2 English for Specific Purposes

From the early 1960's, English for Specific Purposes (ESP) has grown to become one of the most noticeable areas of EFL teaching today (Dudley-Evans 1998). ESP is designed to meet specific needs of the learners and make use of activities on the language appropriation. Knight et al. (2010, p. 7) further clarify ESP when they divide language learners who need ESP into two categories: (1) Language learners who are in the process of developing expertise in their fields need English communication skills as tools in their training. (2) Language learners who are already experts in their fields need English communication skills as tools in their work. ESP courses are designed based on general needs of the learners. The linguist, Anthony (1997) defined ESP as:

> Apparent variations in the interpretation of ESP definitions can be observed; Some people described ESP as simply being the teaching of English for any purpose that could be specified. Others, however, were more precise describing it as the teaching of English used in academic studies or the teaching of English for vocational or professional purposes.
> Anthony (1997: 9–10).

ESP should be designed to provide students with opportunities and resources to pursue their interest in business content and business communication skills in English, to improve their performances in the classroom, and to prepare themselves for their business internships and careers (Knight 2010). Language instructors provides EFL learners meaningful spoken language practice in a class; linking the class to meet their general needs. ESP helps to create a target environment in which language instructors not only offer encouragement and support, but also preparation to learners' future career. Kang and Dennis (1995) stated any attempt to treat English language learning as learning of isolated facts certainly will not promote real knowledge. Students need to learn their target language in context and with visual clues to help them comprehend. The existence of ESP driven course makes language development conceivable. The importance of ESP is its potential to create new opportunities for curriculum and instruction by bringing real-world instruction into the classroom for students to adopt.

Basturkmen (2006) stated the existence of ESP: (1) To reveal subject-specific language use. (2) To develop target performance competencies. (3) To teach underlying knowledge. (4) To develop strategic competence. (5) To foster critical awareness (p. 133). ESP knowledge-based education where learning is rapidly becoming a vocation centered process.

2.3   Technology Assists L2 Learning

Technology provides potential advantages for L2 educational flexible access. The use of the technology helps to decrease the need for on-site teaching accommodation for educators toward students. The role of technology assists learning and autonomous development. Both L2 students and instructors benefit from the support of technology. Paralm et al. (2015) claimed students preferred interactive animations and videos to aid in their learning, furthermore, they also demonstrated increased acceptance of technology assists learning. Additionally, L2 students were allowed to increase control over the pace and timing of the presented material. Bermejo (2005) reported technology-assisted learning enables students produce effective learning environments and facilitates learners to develop appropriate strategies for acquiring and internalizing targeted information and knowledge. Furthermore, technology links the class to the real world, using tools for learners to ask questions and understand the context. Technology offers immediate practice and provides added practice when necessary (Chen 2018). The implementation of innovative technology assisted learning has become an increasing necessary in higher education, deeply influences all aspects of learning. The key elements of technology assist L2 learning are accessibility, immediacy, interactivity and situating of instructional activities (Ogata and Yano 2005). Technology enables L2 teachers to differentiate instruction and adapt classroom activities and homework assignments, thus helping L2 learners enhance the language learning experience (Chen 2018). Technology provides a platform for learners to do meaningful language practice in and after class.

3   Methodology

Design of AR Learning System

There are many consideration issues in designing the AR learning platform such as system complexity, cost, safety, comfort, immersive perception, system adaption and stability. In this paper, to achieve the goal of high immersive experience while keep the hardware cost very low, the google cardboard like HMD structure and a smart phone are combined as the AR display platform. In this way, every school can easily provide the AR learning platform to all the students by offering them the cheap cardboard structure and installing the developed software into their smartphone, while preserving moderate immersive effect than other solutions.

To develop the AR system software, the Unity is adopted as the software development platform. The Unity is a popular and famous game engine with powerful ability in dealing with 3D scene and objects, as well as supporting various AR hardware and software. Moreover, it is easily to program the control UI under Unity environment for the user to interact with the AR system to adjust the digital contents when learning.

On the other hand, to develop the AR learning contents, considering the high authorization fee and the support flexibility for various devices, the Vuforia kit is selected as our development tool since it is free and also fully compatible with the Unity engine and supports the IOS and Android smartphone. In this way, it is easily to develop the AR learning contents and applied to most of the learners' smartphone.

Thus, enables EFL learner to experience the designed AR scene content in the classroom, with high immersive stereo perception, thus can greatly enhancing their learning motivation and effectiveness. The proposed system is similar to the previous structure (Lai 2017, 2018). The illustrations of the AR system prototype and some operation illustrations are shown in Figs. 2, 3, 4 and 5.

Fig. 2. Prototype of the proposed learning support

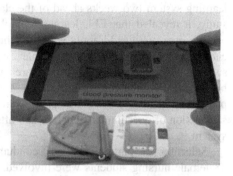

Fig. 3. Illustration of L2 practice between the nursing student and the system.

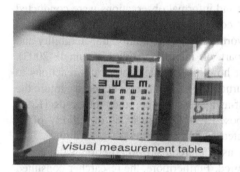

Fig. 4. Illustration of interactive learning system by users.

Fig. 5. Illustration of learning ESP from nursing students.

3.1 Research Data Collection

The purpose of this hermeneutical phenomenological study was to uncover the essence of AR application by the researcher based on ESP context toward nursing students. Hermeneutics is derived from the Greek word hermeneuin, which means "to interpret" (Moran 2000). Students' actual experience of the technology support during the course is widely regarded as providing valuable insights into ESP improvement and is a vital component of a complete instructional evaluation. 48 junior nursing students participated in the study. Students rated these items on a five-point Likert scale that varied from (1 = Strongly disagree, 2 = Disagree, 3 = Neutral, 4 = Agree, 5 = Strongly agree).

Mixed method was applied in the study. The questionnaire of AR assists learning L2 effectiveness and satisfaction was implemented at the first section of the data collection. In addition, the use of semi-structure interview technique was applied to be the main method of gathering qualitative research data. The data was obtained through adapted and modified open-ended questions. Participants were asked to reflect on how AR operated in the ESP context and how AR created the model of interactive learning environments after each AR learning practice.

Participants were assigned to interact with the given scenario of the AR EFL learning system two weeks ahead of the class. According to the researcher's previous study, it revealed that EFL learners complained and claimed that they did not have enough opportunities to apply what they had learn in class. Therefore, the researcher came up an idea of providing flexible English learning environments by using AR support. Students were able to be more familiar with the ESP context and material before attending the course. In addition, students had chances to practice ESP context pre-class and after class. Participants were asked to complete the effectiveness and satisfaction of questionnaires before and after using the AR language learning support system.

Selection of participants was conducted via purposive sampling procedures. 48 female nursing students were involved. The use of semi-structure interview technique was applied to be the main method of gathering qualitative research data. The data were obtained through adapted and modified open-ended questions. Participants were asked to reflect after their AR learning practice at the end of the semester. Individual and group interviews, classroom and informal observations were conducted for an entire semester in order to identify concerns and potential of learning effectiveness. The criteria for establishing trustworthiness are credibility, transferability and dependability. Credibility is the counterpart of internal validity (Connell 2003). Therefore, the researcher must be aware of how his/her interactions and reactions to events affect analysis of the data. A field journal is used by the researcher to capture the interactions and reactions to maintain self-awareness as the study proceeds (Koch 1994; Connell 2003). Creswell et al. proposed (2011) three types of validity of an instrument: content validity, criterion-related validity, and construct validity. The content validity was selected and viewed as a prerequisite to criterion validity for indicating whether the desired trait is measured. Furthermore, the researcher consulted two instructors in English related programs and two instructors from the department of communication engineering to ensure all the open-ended questions covered the research scope. Additionally, after transcribing the qualitative data, two English language instructors verified each transcription.

4 Experimental Results and Discussions

4.1 Quantitative Data

Areas Where Participants' Self-efficacy is Investigated
The first category of question on the survey dealt with the areas in which AR Learning Support System enhance students' self-efficacy. Participants were evaluated on three characteristics of self-efficacy toward the AR learning system. Table 1 provides the

mean ratings of the results. Students feel confident and satisfied using the technological learning system was rated the highest (M = 4.21). Students feel confident and satisfied operating technological learning functions was rated as (M = 4.12). Using the technological system students' daily life was rated as (M = 4.01). All areas were given ratings of above 4.0, students' self-efficacy is positive.

Table 1. Students perceived self-efficacy from the AR technological learning system. The Mean and Standard Deviation (from which means "strongly disagree" to which means "strongly agree").

Items	M	SD
Perceived self-efficacy:		
I feel confident and satisfied using the technological learning system	4.21	0.87
I feel confident and satisfied operating technological learning functions	4.12	0.9
I feel confident and satisfied using the learning contents in my daily life	4.01	0.96

Areas Where Participants' Learning Motivation is Investigated

The second category of question on the survey dealt with the areas in which AR Learning Support System enhance students' learning motivation. Participants were evaluated on three characteristics of learning motivation toward the AR learning system. Table 2 provides the mean ratings of students the results. Students were motivated to use the content of the technological learning system to assist my learning was rated the highest (M = 4.23). Students were motivated to use technological learning system to assist their learning was rated as (M = 4.17). Students were motivated to use the technological learning system as an autonomous learning tool was rated as (M = 4.08). All areas were given ratings of above 4.0, students' self-motivation is high.

Table 2. Students' perceived learning motivation from the AR technological learning system. The Mean and Standard Deviation (from which means "strongly disagree" to which means "strongly agree").

Items	M	SD
Perceived learning motivation:		
I am motivated to use technological learning system to assist my learning	4.17	0.88
I am motivated to use the content of the technological learning system to assist my learning	4.23	0.86
I am motivated to use the technological learning system as an autonomous learning tool	4.08	0.84

Areas Where Participants' L2 Learning Effectiveness is Investigated

The third category of items on the L2 learning effectiveness toward the AR learning system. Three areas in which the system develops learning effectiveness. Item 1 presents the highest agreement of believing the technological learning system can assist L2 learning efficiency (M = 4.15). Students believe AR technological learning system can assist L2 learning motivation has the second highest ratings (M = 4.02). Item 3 investigates AR technological learning system can assist L2 learning performance (M = 3.93). All areas were given ratings of above 3.90. Students' L2 learning effectiveness is developed by the system (Table 3).

Table 3. Students' learning effectiveness

Items	M	SD
L2 Leaning effectiveness:		
I believe AR technological learning system can assist L2 leaning efficiency	4.15	0.86
I believe AR technological learning system can assist L2 learning performance	3.93	0.80
I believe AR technological learning system can assist L2 learning motivation	4.02	0.93

4.2   Qualitative Data

Research Question 1
How are the Nursing students' perceptions of how AR facilitates ESP Learning?
Technology facilities learning takes its own pace and extent to show the world how best to employ various forms of digital technology by cooperating with AR to prepare students in the effective use of technology for language learning and for digital literacy. According to students, L2 learning system was dependent on content, process, teacher and student characteristics.

"AR has given rise to significant changes to my language learning; I felt like I am in the real working environment. The expansion of technology advantages the traditional instruction. English learning becomes more entertained and motivated with AR implementation". (May, female student)

Regarding AR facilitates ESP instructions, nursing students expressed that teaching should be connected to their current level of knowledge and skills. Content should be prioritized according to practical relevance.

"After receiving theories from the textbook, the instructor applies AR to provide activities and tasks to integrate my course progress and assess my English skill. The language learning classroom becomes an energetic environment where I feel comfort and confident to speak and repeat what I learned and heard." (February, female student)

Students addressed that a strong focus on pedagogy specific to ESP, feedback from the instructor is essential to facilitate improvement in learning outcome.

"AR enables me to express my views on courses. Right after interacting with AR activities, I like the instructor to provide specific feedback to my practice and tell me if I meet the requirement of the course." (April, female student)

When the course instruction sets the stage for the Technology use, it provides a rationale and overview for the rest of the course; students are willing to participate in the class, interact with partners, and provide positive evaluation for language teachers.

"I would give a positive rating for this course in which I always have learned a lot in a joyful environment and pleasant manner. My classmates and I enjoy having AR to facilitate our learning progress. I am able to say new vocabulary out loud, practice the conversation drills, and enhance my pronunciation. (June, female student)

Nursing students expressed their opinions on teachers' attitude and profession; they anticipate their teachers to be knowledgeable, clinically experienced, well-prepared, and enthusiastic.

"I like dedicated teachers who view educational activities as an integral part of their professional role. For example, teachers should be open and knowledgeable toward new information and technology. (July, female student)

Research Question 2

How are the nursing students' perceptions of how AR facilitates their behavior of learning?

ESP learners are provided with an opportunity to engage in authentic practices related to their areas of study and work. Students were well aware of their own responsibility for achieving their learning outcomes.

"I used to anticipate the answers for the English practice from my language instructor but after AR facilitates the instruction and my learning, I become more independent and braver to try to speak English and practice new lessons. AR provides me the environment to practice and apply the target language before and after the class." (September, female student)

A lack of motivation in learning was mentioned from two of the participants. After AR implement, students showed their concerns and changed their attitude in learning.

"I think the motivation in learning English is very important, I used to feel bored in class because I don't like to just sit and listen to the lecture. I think I can do self-study at home without coming to school. After practice English by AR technology, learning becomes real and practical". (December, female student)

Students embraced AR with full participation, they expressed that they had more freedom and choice in learning. Opportunities of learning the target language become creative and innovative. The involvement of technology in a classroom lessens diminish disruptions in the learning environment. Technology integrated into practical lessons, students act out less because using AR is one of their strengths.

"This course is valued. When I put on the cardboard, I become another person. I have no fear to ask my teacher anything in English and she'll respond. I used to speak English with the voice that no one could hear but not anymore. Applying what I learned in the daily life and work place is fun and stress free." (January, female student)

A key element in the academic success of students is active engagement in their learning. The researcher observed a big change in how her students were engaged in content and activities during the course.

"The instructor walked around and checked with the students whether they need assistance or not. Nursing students were more actively taking on their learning. They showed their participation. They are all engaged in the learning and taking it seriously. With the AR support, they're completely excited. There is so much more motivation. Some of the students resisted to return the 3D glasses and stated to keep them for learning." (Field Note)

5 Conclusions

The results indicated that implementing AR was a vital strategy for producing independent thinkers and learners. According to Chen (2018) technology offers immediate practice and provides added practice when necessary. Technology flourishes under ESP-driven and creative approach, nursing students will be able to develop their own inquiries. Effective interaction improves the language acquisition process and motivates EFL learning. Technology helps to create an active environment and promote verbal communication. Interactive instructional support system plays a beneficial role in facilitating nursing students to reduce obstacles and learn through technological input alone which reflect to the theory of providing nursing students to learn in a target scenario based simulation (McAllister et al. 2013). AR was considering to facilitate ESP instruction due to various limitations of physical instruction in nursing education. AR allows educators to take advantage of these learning experiences that happen around us and bridge the gap between where we learn language and where we employ our new skills. Technology enables L2 teachers to differentiate instruction and adapt classroom activities and homework assignments, thus helping L2 learners enhance the language learning experience (Chen 2018). AR assists language learning may not yet be currently reflected in the curricula of English instruction, the evidence of interest from the nursing student in this study frames a scratch for educational technology developers. Overall, much more research about AR offers a promising way to mediate learning opportunities and facilitates teaching instruction and learning motivation is required in order to distinguish the field for constructing new conceptual and theoretical teaching models.

References

Alsied, S.M., Pathan, M.M.: The use of computer technology in EFL classroom: advantages and implications. IJ-ELTS **1**(1), 61–71 (2013)

Anthony, L.: ESP: What does it mean? ON CUE (1997). http://interserver.miyazaki-med.ac.jp/~cue/pc/anthony.htm

Basturkmen, H.: Ideas and Options in English for Specific Purposes. ESL and Applied Linguistic Professional Series. Eli Hinkel, London and New Jersey (2006)

Bermejo, S.: Cooperative electronic learning in virtual laboratories through forums. IEEE Trans. Educ. **48**(1), 140–149 (2005)

Chen, Y.: Reducing language speaking anxiety among adult EFL learners with interactive holographic learning support system. In: Wu, T.-T., Huang, Y.-M., Shadieva, R., Lin, L., Starčič, A.I. (eds.) ICITL 2018. LNCS, vol. 11003, pp. 101–110. Springer, Cham (2018). https://doi.org/10.1007/978-3-319-99737-7_10

Connell, P.J.: A phenomenological study of the lived experiences of adult caregiving daughters and their elderly mothers. University of Florida (2003)

Creswell, J.W., Plano Clark, V.L.: Designing and conducting mixed Methods research. Sage Publications, Thousand Oaks (2011)

Dudley-Evans, T.: Developments in English for Specific Purposes: A Multi-Disciplinary Approach. Cambridge University Press, Cambridge (1998)

Hatch, E. (ed.): Second Language Acquisition: A Book of Readings. Newbury House, Rowley (1978)

Hewitt-Taylr, J.: Technology-assisted learning. J. Furth. High. Educ. 27(4), 457–464 (2010)

Hughes, C.E., Stapleton, C.B., Hughes, D.E., Smith, E.M.: Mixed reality in education, entertainment, and training. IEEE Comput. Graph. Appl. 25(6), 24–30 (2005)

Koch, P., Oesterreicher, W.: Schriftlichkeit and Sprache. In: Gunther, G., Otto, L. (eds.) (1994)

Knight, K.: English for specific purposes (ESP) modules in the self-access learning center (SALC) for success in the global workplace. Stud. Self-Access Learn. J. 1(2), 119–128 (2010)

Knight, K., Lomperis, A., van Naerssen, M., Westerfield, K.: English for specific purposes: an overview for practitioners and clients (academic and corporate). PowerPoint Presentation Submitted to Alexandria, Virginia: TESOL Resource Center (2010). http://www.tesol.org/s_ tesol/trc/uploads/Other/119485/1564_Knight_ESPPPTforTRC.pdf

Lam, Y.: Technophilia vs. technophobia: a preliminary look at why second-language teachers do or do not use technology in their classrooms. Can. Mod. Lang. Rev. 56(3), 389–420 (2000). http://dx.doi.org/10.3138/cmlr.56.3.389

Long, M.: The role of the linguistic environment in second language acquisition. In: Ritchie, W. C., Bathia, T.K. (eds.) Handbook of Research on Second Language, pp. 413–468. Academic Press, San Diego (1996)

McAllister, M., et al.: Snapshots of simulation: Creative strategies used by Australian educators to enhance simulation learning experiences for nursing students. Nurse Educ. Pract. 13(6), 567–572 (2013)

Mankin, D., Cohen, S., Bikson, T.: Teams and technology: tensions in participatory design. Organ. Dyn. 2616(97), 63–76 (1997)

Milgram, P.: A taxonomy of mixed reality visual displays. IEICE Trans. Inf. Syst. E77-D(12) (1996)

Ogata, H., Yano, Y.: Knowledge awareness for computer-assisted language learning using handhelds. Int. J. Learn. Technol. 5(1), 435–449 (2005)

Paralm, M., Shenoy, P., Loh, K.Y.: Students' perception of technology-assisted learning in undergraduate medical education-a survey. Soc. Sci. J. 52(1), 78–82 (2015)

Moran, D.: Introduction to Phenomenology. Routledge, Taylor & Francis Group, New York (2000)

Richards, J.C.: Foreword. In: Agudo, J.D.D.M. (ed.) English as a Foreign Language Teacher Education: Current Perspectives and Challenges, pp. 1–3. Rodopi, Amsterdam (2014)

Kang, S.H., Dennis, J.R.: The effects of computer-enhanced vocabulary lessons on achievement of ESL grade school children. Comput. Sch. 11(3), 25–35 (1995)

Lai, C., Chu, Y.: Increasing the learning performance via augmented reality technology. In: Wu, T.-T., Gennari, R., Huang, Y.-M., Xie, H., Cao, Y. (eds.) SETE 2016. LNCS, vol. 10108, pp. 58–64. Springer, Cham (2017). https://doi.org/10.1007/978-3-319-52836-6_8

Lai, C., Chang, Y.: Improving the skills training by mixed reality simulation learning. In: Wu, T.-T., Huang, Y.-M., Shadieva, R., Lin, L., Starčič, A.I. (eds.) ICITL 2018. LNCS, vol. 11003, pp. 18–27. Springer, Cham (2018). https://doi.org/10.1007/978-3-319-99737-7_2

Improving Supply Chain Resilience
with a Hybrid System Architecture

Yu Cui[1](✉), Hiroki Idota[2], and Masaharu Ota[3]

[1] Graduate School of Business Administration and Economics, Otemon Gakuin
University, Osaka, Japan
yucui@otemon.ac.jp
[2] Department of Public Management/Graduate School of Economics,
Kindai University, Osaka, Japan
[3] Faculty of Business Administration, Osaka Gakuin University, Osaka, Japan

Abstract. The purpose of this research is to propose a new hybrid system architecture, which is a solution for reducing and even avoiding the complexity and vulnerability of the whole supply chain. In this paper, we propose a conceptional model which introduces some features of decentralized system architecture based on blockchain technologies, such as new consensus mechanism and smart contract into a typical supply chain system so that defects of traditional centralized system will be solved and supply chain resilience can be improved. Furthermore, some fatal flaws of current blockchain system can be shored up in an IoT infrastructure environment which is consolidating supply chain resilience. Last but not lease, we identify some unsettled issues and drawbacks derived from the hybrid system architecture and indicate the solutions regarding these challenges.

Keywords: Supply chain resilience · Hybrid system architecture · Blockchain

1 Introduction

Supply chain is a network chain structure that connects suppliers, manufacturers, transporters, retailers and consumers as a whole [1]. Nowadays, with the construction of supply chain system, more and more enterprises realize supply chain informatization and meet the business needs of entities in supply chain, meanwhile, the operational processes and information system cooperate closely, accomplishing the sharing of information among participating entities and the seamless connection of each link process. By doing so, the entities can form an interest community through supply chain, and the incentive to strengthen accountability and supervision arises. Along with this, the overall automated production efficiency and the profit will be improved, while the cost will be reduced accordingly.

In the meantime, with the continuous promotion and application of advanced technologies such as Internet of things (IoT), cloud computing, big data, Artificial Intelligence (AI), robotics, and radio frequency identification, not only the entire information sharing from upstream suppliers to downstream vendors is enhanced, but also the level of lean management of supply chain is promoted [2].

© Springer Nature Singapore Pte Ltd. 2019
J. C.-W. Lin et al. (Eds.): MISNC 2019, CCIS 1131, pp. 142–154, 2019.
https://doi.org/10.1007/978-981-15-1758-7_12

On the other hand, the core of the concept of supply chain is to build trust among various entities. Through product design, raw material procurement, production, transportation and sales, the relevant upstream and downstream enterprises on all links of the transaction are chained together, and the information collection on the chain is integrated. Although more and more enterprises are involved in supply chain, the information sharing among different enterprises is limited, and the complete information on the entire chain cannot be symmetric and transparent, accordingly, there may be risks of false information and tampering with historical information.

Moreover, in recent years, remarkable achievements have been made in the industry of Internet of things, and a large number of sensors and machinery equipment are integrated with the Internet, resulting in intelligent management and operation. Gubbi et al. (2013) indicated a scenario that reflects the application of Internet of things throughout the world in future. In particular, all intelligent devices and objects are embedded with sensors. Through real-time perception of the surrounding environment and network interconnection, it is possible for people to complete work, business, transactions and etc. without leaving home [3]. Nevertheless, with the development of Internet of things, device security, bottlenecks such as device security, privacy protection, structural rigidity, communication compatibility and multi-agent collaboration still need to be broken through.

2   Drawbacks Arising from the Applications of IoT in Supply Chain

From the following aspects, we can roughly sum up the drawbacks arising from the application of Internet of things in supply chain (see Fig. 1):

(1) Architecture rigidity
   At present, almost all Internet of things applications adopt centralized architecture, and Internet of things operators provide data storage and exchange services for Internet of things applications through cloud server cluster. With the development of low-power WAN technology, Internet of things terminals will grow exponentially in future [4].

(2) Equipment security
   The distributed denial of service attack launched by the massive Internet of things devices controlled by it has caused the instant paralysis of an American network domain name resolution service provider called Dyn [5].

(3) Communication compatibility
   There are multiple competitive standards and platforms in the field of Internet of things communication. Furthermore, there are still some problems in the communication compatibility of IoT device groups using different communication means [6].

(4) Privacy protection
   Traditional Internet of things mainly adopts a centralized management architecture, and there are many shortcomings in the aspects of user privacy and secure channel transmission [7].

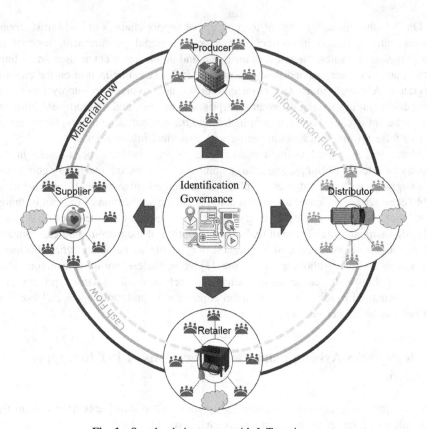

Fig. 1. Supply chain system with IoT environment

To solve the above problems, a decentralized architectural system based on block chain and smart contract is designed in this study. Through blockchain technology, the system keeps complete records of all the data throughout the processes of raw material supply, product manufacturing and processing, logistics transportation and final sales of suppliers, manufacturers, transporters, retailers and other supply chain agents. The recorded data is transparent, traceable and unalterable, and the privacy data is encrypted, resolving the trust issue among supply chain entities, regulatory traceability problem and data privacy protection among the main entities of supply chain.

Meanwhile, based on the continuation of the centralized system framework, Internet of things technologies such as QR code, Radio Frequency Identification (RFID) and short-range wireless communication technology are adopted to maximize the potential of the existing centralized system framework and realize the intelligence of information collection, enabling the perfect integration of centralization and decentralization architecture in improving the overall efficiency of the supply chain system, in the meantime, a genuine resilient supply chain is established as the seamless connection between business and nodes.

3   The Principle of Blockchain

3.1   The Definition of Blockchain

In 2008, the paper "Bitcoin: A Peer-to-Peer Electronic Cash System" published by Satoshi Nakamoto brought Bitcoin into public's view [8]. In this paper, a new type of electronic cash system is proposed, which is based on cryptographic principles rather than the traditional information system. Any agreed parties can make transactions without the participation of third-parties. And bitcoin can continue to operate stably without the operation or management of any centralized organization. As time goes by, people gradually come to realize the significance of the underlying technology of Bitcoin, and extract it to define as blockchain technology.

Blockchain technology and its application is currently a topic in the spotlight. It generally refers to the application platform based on blockchain technology. Each functional entity of the blockchain (including blockchain application and blockchain service platform) jointly maintains one or more distributed peer-to-peer ledgers, and the data in the distributed ledger is encrypted and stored, making it hard to be rewritten or forged [9]. Blockchain technology is highly anticipated and considered to be a trusted, accountable, transparent and efficient trading application [10]. It can be applied in the fields of financial services, healthcare, government, manufacturing, retail, media and entertainment, supply chain and logistics.

Blockchain technology builds a trustworthy decentralized system through distributed consensus algorithms and digital encryption, accomplishes information transfer and value transfer synchronously, and reaches consensus and establishes mutual trust mechanism in distributed nodes [11]. Through the utilization of cloud services and Security Keeper, information resources such as smart government and think tank are integrated to provide relatively neutral platform for members of blockchain, meanwhile, transparency of data transmission is enhanced; regulatory functions and social supervision functions of government can be fully exerted, and diversity of information resource accesses can be realized.

3.2   The Characteristics of Blockchain

Blockchain is a kind of distributed ledger that collectively maintains a reliable database technology solution through decentralization and trustless [12]. In this scheme, blocks are linked orderly to form a chained data structure, forming a distributed database that can hardly be changed. The "distributed" here does not only reflect the distributed storage of data, but also reflects the distributed recording of data.

Each node on the blockchain stores all the information of the whole blockchain, and each node in the chain can be understood as the backup node of the whole data. The data of each node is owned, managed and supervised by all participants, and all participants can update the database and confirm it [13]. Decentralization, autonomy, openness and non-usurpation are the core advantages and characteristics of blockchain, which can effectively reduce the cost of trust between participants.

The blockchain technology does not require central scheduling, meanwhile, the authenticity of data can be effectively guaranteed, data collaboration of all participants in supply chain and the construction of digital trust system can be promoted, realizing a new management architecture between upstream and downstream enterprise [14]. The chained account structure formed by blockchain greatly enhances the verification and storage performance of data.

Network programming to achieve business logic. Both distributed algorithms and cryptographic signatures are used in implementing point-to-point networks. Node search, data transmission, verification, and consensus algorithms all belong to this layer.

In blockchain technology, a smart contract is made through the participation of various users. With programming language, this is implemented and publicized to each node. Once the transaction is completed jointly, the transaction status is saved into blockchain through consensus algorithm. The adoption of consensus algorithms and cryptographic algorithms is to maintain the consistency of nodes and the security of data transmission and access [15]. A smart contract contains the rights and obligations of both parties, the triggered condition, and the actions that are performed after the trigger.

3.3   Blockchain Classification

According to the scope of application of the blockchain and node permissions, it can be divided into three categories.

I.   Public Blockchain: The interior of the public blockchain is completely decentralized. Each node of the distributed system can freely access the network and participate in the process of reading, writing, verifying and consensus of data on the chain. This is the earliest and currently the most widely used blockchain, and anyone can participate in its consensus process, such as Bitcoin based on the blockchain.

II. Private Blockchain: is a completely centralized structure. The access permission of each node of the distributed system is subject to a central organization, and the access permission can be selectively opened depending on the situation. The private blockchain only uses the general ledger technology of the blockchain to record accounts. It can be a company or an individual, and enjoys the access permission of the blockchain.

III. Alliance Blockchain is a partially decentralized structure. Each node of the distributed system needs to be authorized by the pre-selected node to access the network, and only the pre-selected nodes can participate in the consensus process. A plurality of pre-selected nodes are designated as billers within a certain group, other access nodes can participate in the transaction, but not check the accounting process.

4   Introducing the Blockchain System Under a Decentralized Framework

Based on the blockchain system, issues such as digital identity, data assets that are related to the interconnection, trusted sharing and orderly collaboration of private data can be effectively solved, enhancing the collaborative efficiency of business internet and reducing the collaborative cost. success. Accordingly, the multi-level transmission of credit, capital, business and etc. is realized. Therefore, blockchain system is an important basis for the construction of collaborative ecology suitable for the new business environment.

In summary, this paper proposes to utilize blockchain technology to build a new framework in supply chain in which blockchain system is the core, smart contracts, authentication centers, and public inquiries are practical application medium. In this way, all the participants can check the progress of the supply chain. Meanwhile, smart contracts ensure that transactions are traceable and transaction information is transparent, averting contract frauds and ensuring the security and control of the transactions.

4.1 The Blockchain System

Blockchain is a continuously growing distributed database that is jointly maintained by multiple parties. The key is to establish the trust relationship between each other through distributed network, chronological cryptography and distributed consensus mechanism, and a trusted mechanism for information transmission and exchange in untrusted networks is established.

In terms of transaction processing, different from the ACID principle utilized in the traditional database, blockchain adopts the final consistency principle so as to ensure that all the data of nodes can finally be consistent after a period of synchronization. The characteristics of decentralization, common maintenance, data transparency, tamper resistance and etc. of blockchain enable it to better suit the supply chain that has multiple participants and in which an open, transparent and mutual trusted mechanism is essential.

Blockchain also speeds up the response to emergencies. Once problem occurs, blockchain can trace the entire process of production, find out the stage where problem occurs, and make a decision on whether to recall it within a few minutes. In the meantime, in the recall process of industrial products, the shared account book of blockchain can specifically identify the specific batch where the problem occurs, and avoid extensive recall.

In addition, because blockchain links isolated data of various enterprises, there will be more data sources for big data based on the supply chain, which will greatly increase the inventory and quality of the data and make it possible for big data to be used better. At the same time, as blockchain data cannot be tampered, the credibility of data makes it possible for enterprises to rely on data, promoting the establishment and prosperity of the big data market accordingly.

The blockchain includes a block header and a block body. The hash value of the block header is utilized as the unique identity of the block in blockchain and is stored in the "hash value of the previous block" field of the block header of the next block. By recording the hash value of the block header, a data chain that can be traced from the latest block to the first block is formed, ensuring that the supply chain system can trace the data of each subject at each time stage.

Since each block is linked with the previous block by means of cryptography, after blockchain reaches a certain length, to modify the transaction content in a certain historical block, it is necessary to reconstruct the transaction records and cryptographic proofs of all blocks. And tamper-resistance can be effectively realized by doing so. In data storage layer, block structures, digital signatures, hash functions, Merkel trees, and asymmetric encryption techniques are involved.

In addition, the "hash value of data" field of the block body records the hash value of each data of each subject in the time phase of the block. Therefore, each data of each subject in the supply chain system can be traced back. Through these two parts of the blockchain, the data of the supply chain system can be traced back and it assists the accountability and supervision.

The storage process of block chain system is to store the newly generated data of each entity into the newly generated block, and ensure that the data of each entity as the node's own server is consistent (see Fig. 1). In detail, the specific process is as follows: (1) the blockchain system uses the consensus mechanism to select the authorized nodes among all the nodes of the supply chain; (2) the new data generated by the nodes of each entity is transmitted to the authorized nodes through point-to-point transmission; (3) the authorized node uses the consensus mechanism to generate new blocks periodically; (4) the authorized node processes the new data using signature algorithm, hash algorithm, and etc., and adds fields such as "hash value of the previous block", "timestamp" and "difficulty value" and so on, and fills these into the new block; (5) the authorized node uses the point-to-point transmission mechanism to transmit the new block to the whole network; (6) after receiving the new block, each node adopts the signature algorithm, hash algorithm, and etc. to verity, and once the verification succeeds, the new block is added to the end of the existing blockchain.

The storage process of blockchain system utilizes technologies such as consensus mechanism, point-to-point transmission, signature algorithm and hash algorithm to ensure the consistency, tamper protection and security of data, so the data of each node is open and transparent to other nodes, raising the trust among the main entities of the supply chain. In the figure below, the storage process of the blockchain system is illustrated.

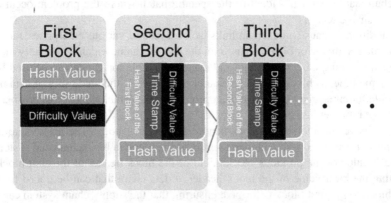

Fig. 2. The image of new block in the blockchain system

4.2   Smart Contract System

A smart contract is a contract that record provisions in computer language rather than legal language. It can be viewed as a program that can be automatically run in a blockchain, and it covers programming languages, compilers, virtual machines, events, state machines, fault-tolerant mechanisms, and so on.

The smart contract system in supply chain applies the smart contract and the contract code based on block chain technology. The execution process and result are open and transparent to all the entities of the supply chain, meanwhile, the result cannot be tampered with, enhancing the credibility of the supply chain, and it is beneficial for the supervision and traceability.

The smart contract system is responsible for providing the interactive interface including the two subsystems called contract generation and contract execution. The contract generation subsystem submits the code of the smart contract to the blockchain system for storage. The contract execution subsystem is responsible for running the code of the smart contract and realizing the storage or query of data of the blockchain system.

4.3   Authentication Center System

The authentication center system is mainly in charge of the generation of public and private keys, enabling the generation and authentication of new blocks for the Internet of Things and the blockchain system respectively. To enhance the security of the system, the public key and private key will be updated regularly, and a new key version number will be generated each time. In the meantime, the key version number will also be forwarded to the Internet of things and the blockchain system.

The encryption and decryption process of the private data is as follows: after the authentication center generates the public key, the private key and the key version number, the public key and the key version number are sent to the Internet of things system. In the meantime, the private key and the key version number are sent offline to the supervision subsystem of the inquiry system; the Internet of things system adopts the public key to encrypt the privacy data, and the privacy data and the key version number are uploaded through the smart contract system after encryption.

Through the supervision subsystem of the inquiry system, the department in charge can decrypt the corresponding private key with the key version number and check the private data. The process is shown as below.

4.4   Public Inquiry System

The public inquiry system is mainly constructed for the inquiries from various entities and consumers in the supply chain. Through the adoption of smart contract system, the system queries the data from the blockchain system and provides it to suppliers, manufacturers, transporters, retailers and consumers.

The entities and consumers in the supply chain utilize the public inquiry system to query public data such as supplier data, transporter data, manufacturer data and etc. by the interfaces in the smart contract module.

5 Building a Hybrid System Architecture Under IoT Infrastructure Environment

On the other hand, although blockchain technology has a broad prospect of application in the supply chain system, there are still a variety of constraints and obstacles in its development (Fig. 3).

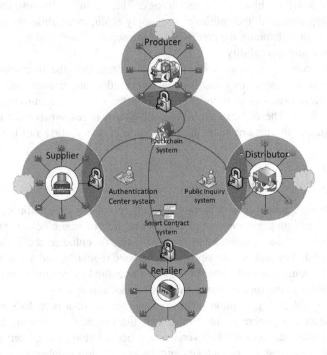

Fig. 3. The diagram of theoretical hybrid system architecture

(1) Resource consumption:
   The equipment of current block chain system generally has problems such as low computing power, weak networking capabilities and so on. The Proof of Work (PoW) adopted by Public Blockchain, which is represented by Bitcoin, consumes numerous computing and power resources, and is not applicable to the supply chain system environment. Developing the swarm intelligence that applies an effective interaction mechanism to aggregate and utilize the distributed consensus nodes is an important issue that needs to be solved urgently in block chain combined within supply chain.
(2) Bottleneck of performance:
   Blockchain requires each node in the system to keep a backup of the data, which is extremely unrealistic for the rapidly growing mass data storage of the supply chain. Although some of the problems can be solved with lightweight authentication nodes, the industrial-scale solutions for larger scales remain to be studied and

designed. The traditional blockchain can only process seven transactions per second, plus the consensus confirmation time, it takes about an hour to write to the block. Therefore, it is not suitable for modern supply chain systems that are delay-sensitive, such as industrial internet.

(3) Partition tolerance:

Since the equipment of modern supply chain system may be placed in depopulated zone, complex electromagnetic area, fast mobile situation and etc., issues such as invalidation of nodes, frequent joining or exiting the network and etc. may occur. This will consume large amount of network bandwidth, resulting in network instability and even network fragmentation. More serious challenges are posed to the effective operation of the blockchain system.

However, with the development of IoT fog computing and edge computing, in the current supply chain system based on the centralized framework, IoT technologies such as RFID and Near Field Communication (NFC) technology are fully utilized. Meanwhile, the blockchain technology represented by consensus mechanism and smart contract is introduced, so as to construct a new type of supply chain system framework in which centralization and decentralization coexist and complement each other.

To sum up, based on the introduction of the blockchain structure under the decentralized framework proposed above, this paper reconstructs the supply chain under the centralized framework and integrates the IoT system into the blockchain architecture, so the problems caused by the blockchain architecture can be overcome and resolved.

What are realized through IoT system include: entities of the supply chain system such as suppliers, manufacturers, transporters, retailers and so on, according to their own characteristics, after selecting a scanning device of two-dimensional code, RFID, NFC to scan the product, the system will automatically obtain the information of the product, including the product name, product model and the like.

And then, the product information and the relevant data submitted by the entities of the supply chain, including the company name, company address, source of raw materials, time of manufacturing and processing, location of manufacturing and processing, start time and arrival time of transportation, location of sales, time to market and other public data. Furthermore, private data such as legal representatives of the company, internal model No. of the product, internal data of the product, tax payment detail and etc. that have been encrypted by the public key of the authentication system, are uploaded and saved to the block chain system through the interface provided by the smart contract system as a data source for the entire supply chain.

The system obtains product information by scanning two-dimensional code, RFID and NFC, and the entities of the system submit relevant data, including public data such as company name, company address and randomly generated privacy data. After the encryption of private data with public key, public data and privacy data are stored together in the blockchain module through the smart contract module. The public inquiry sub-module of the supply chain module can inquiry public data through the smart contract module, and the data obtained from the query are completely consistent with the data obtained after scanning.

The supervision sub-module of the supply chain module can decrypt the privacy data through the adoption of the private key, and the privacy data obtained by query is completely consistent with the one submitted to the system. Through this prototype system, operations such as uploading, storing, querying, encrypting and decrypting the entire supply chain data are realized. The data is transparent, traceable and irrevocable, and the problems of trust, regulatory traceability, protection of data privacy and automation among the entities are solved.

In accordance with the technical characteristics of blockchain and IoT applications, this proposal can effectively resolve the issues including the identity proof of participants, the confirmation of digital assets and the unification of transaction rules. The supply chain of enterprise based on blockchain technology collaborates with upstream and downstream enterprises to complete the transaction, and multi-party collaborative work assists the formation of a more transparent, intelligent, and real industrial ecological chain. Due to the synergy effect of blockchain technology, business agreement can be executed justly and automatically without interference (see Fig. 2).

Meanwhile, the digitization of assets entitles the value of assets the characteristics of certainty, transferability and separability, and effectively enhances the liquidity of digital assets. As all the transactions are completed with the participation of various parties in the ecological chain, all the parties possess integrated information of transaction, which in turn can guarantee the authenticity of information of contracts and orders. Moreover, the mechanism based on cryptography ensures that the transaction cannot be denied or destroyed. The application of blockchain technology impose strict requirements for the identity proof of nodes in the process of transaction process, resolving the issue of identity authenticity of the participants.

6   Conclusion and Prospect

At the beginning of the Internet, the earliest and most important problem was the creation and transmission of information. People can quickly generate and copy information to every corner of the world through the Internet. However, the Internet still cannot solve the value and credit transfer, and cannot realize the point-to-point delivery of value.

The so-called value transfer means that everyone in the network can accurately transfer a certain part of the value from one address to another by means of recognition and confirmation, and must ensure that when the value is transferred, which means that the original address is reduced and transferred, at the same time, the new address increases the value of the transfer.

In the context of the Internet, how can you prevent malicious destruction, deception, and confusion when you need to exchange value with unfamiliar opponents, thus reducing false decisions? In other words, how do you reach consensus on the nodes that are distributed across the network in the absence of trusted channels and trusted central nodes? Blockchain technology provides the possibility to solve these problems.

In our review of the application of blockchain and smart contracts in academic and practical fields, it is revealed that the content involved in supply chain applications is relatively less than others. Based on its technical advantages, the application of blockchain in various industries has been gradually promoted, which can effectively solve many inextricabilities in the field of supply chain. Applying blockchain and smart contracts into supply chain management is to implement a decentralized framework setting for supply chain management demand through the construction of blockchain system, and to realize the perfect interaction between centralized and decentralized framework.

On the other hand, the application of blockchain technology in the field of IoT has started, and the gradual integration and advancement of blockchain and IoT are the important objectives of next step. Due to its characteristics of peer-to-peer, openness and transparency, multi-party consensus, tamper-resistant of information, and programmable, blockchain will exert a far-reaching impact on the technology system, infrastructure, operation mode and other aspects of the IoT, putting forward new options and choices for solving the problems encountered in the development of supply chain.

I. Reduce the operating costs of supply chain. Blockchain combined with IoT can use P2P transmission to solve the problem of limited network bandwidth, and adopt the computing resources, storage resources and data resources of hundreds of millions of IoT idle nodes in a distributed manner so as to greatly reduce the high operation and maintenance costs of the traditional centralized architecture of the IoT, and promote the flow and sharing of information.

II. Solve the security issue of nodes in different trust domains. Blockchain combined with IoT can adopt the timestamp-based storage structure and distributed storage mechanism to construct a traceable electronic token to protect and ensure the data is integrate and tamper-proof.

III. Realize the intelligent collaboration among the devices of IoT. Smart contracts using blockchain can transform traditional intelligent devices of IoT into independent self-maintaining and self-regulating individuals which are able to exchange information or verify identity with each other based on pre-defined or embedded rules and contracts without manual participation or intervention.

Blockchain technology can promote the interconnection of information among enterprises in the supply chain and increase information transparency. Meanwhile, as the security of individual enterprise information is ensured with mathematical algorithms, so business process can be simplified and risk costs can be reduced. If the blockchain technology can be applied to supply chain resilience appropriately, the efficiency of supply chain management will be further improved.

Based on the introduction of blockchain system and IoT system for supply chain management demands, this paper attempts to make an ideal integration of the centralized and decentralized framework.

Acknowledgments. This research is supported by Grant-in-Aid for scientific Research of the Japanese ministry of Education, Culture, Sports, Science and Technology under the Contract No. (C) 19K01824 (2019–2021).

References

1. Zhang, D., et al.: Process of innovation knowledge increase in supply chain network from the perspective of sustainable development. Ind. Manag. Data Syst. **118**(4), 873–888 (2018)
2. Addo-Tenkorang, R., Helo, P.T.: Big data applications in operations/supply-chain management: a literature review. Comput. Ind. Eng. **101**, 528–543 (2016)
3. Gubbi, J., et al.: Internet of Things (IoT): a vision, architectural elements, and future directions. Future Gener. Comput. Syst. **29**(7), 1645–1660 (2013)
4. Dano, M.: EU's connected continent' proposal draws muted response, but analysts upbeat, Fierce Wireless Europe, p. 1, 9 December 2013
5. Kerner, S.M.: Large DDoS attacks on the rise. Akamai Report Finds, eWeek, p. 1, 16 February 2017
6. Markkanen, A.: Operators rally to NB-IoT as technologies such as Sigfox, Ingenu and LoRa rise. WiMAX Mon. Newsl. **26**(3), 8 (2016)
7. Weber, R.H.: Internet of Things: privacy issues revisited. Comput. Law Secur. Rev. **31**(5), 618–627 (2015)
8. Nakamoto, S.: Bitcoin: A Peer-to-Peer Electronic Cash System (2008). https://bitcoin.org/bitcoin.pdf
9. Swan, M.: Blockchain: Blueprint for a New Economy. O'Reilly Media, Sebastopol (2015)
10. Marsal-Llacuna, M.-L.: Future living framework: is blockchain the next enabling network?' Technological Forecasting and Social Change (2017). In press, corrected proof, Available online 12 December
11. Tapscott, D., Tapscott, A.: Blockchain Revolution: How the Technology Behind Bitcoin is Changing Money, Business, and the World, Portfolio (2016)
12. Yusuf, S.: Blockchain/Distributed ledger technologies: where they came from, where they are heading. Econ. Cult. Hist. Jpn. Spotlight Bimon. **37**(6), 18–22 (2018)
13. Noguchi, Y.: Blockchain Revolution: The Advent of Distributed Autonomous Society. Nikkei Publishing Inc. (2017). (in Japanese)
14. Kshetri, N.: 1 Blockchain's roles in meeting key supply chain management objectives. Int. J. Inf. Manag. **39**, 80–89 (2018)
15. Kshetri, Nir: Blockchain's roles in strengthening cybersecurity and protecting privacy. Telecommun. Policy **41**(10), 1027–1038 (2017)

An Interactive, Location-Aware Taiwanese Social Network for Both Everyday Use and Disaster Management

Tzu-Ping Lai[1], Shun Yao[2], Wing-Lun Siu[3], Yu-Che Cheng[1], Heng-Yi Su[2], and Yi-Chung Chen[1(✉)]

[1] National Yunlin University of Science and Technology, Yunlin, Taiwan, R.O.C.
s292423@gmail.com, npes60715@gmail.com, mitsukoshi901@gmail.com
[2] Feng Chia University, Taichung, Taiwan
yaoshuncy@me.com, hengyisu@fcu.edu.tw
[3] National Cheng Kung University, Tainan, Taiwan
alansiu4389@hotmail.com

Abstract. The concept of leveraging social network crowdsourcing to help governments respond to and manage disaster events has become popular in recent years. However, existing methods in Taiwan have been challenged by the following difficulties. (1) Information crowdsourced from existing social networks has typically been limited by privacy settings (which can prevent data collection) and formatting issues (which can affect data processing). (2) Existing social networks lack location-aware functions, which makes it difficult to do the spatial analyses of disaster events. (3) Platforms which have been created to collect disaster information do not typically attract enough users to facilitate effective crowdsourcing of information following an disaster. To address these limitations, this study proposes a novel, interactive, and location-aware Taiwanese social network. Unlike social networks which are currently popular, our proposed network focuses on geographic space, and users can only view posts which were generated near their current location or close to where they live. Under normal, non-disaster circumstances, users can employ our proposed social network to view issues in their surrounding environment. In the event of an disaster, users can employ our social network to provide disaster management agencies with comprehensive information about the disaster. In addition, disaster management agencies can use the proposed social network to interact with users who are located near the disaster. We anticipate that our proposed social network will better facilitate information crowdsourcing than convention social networks in Taiwan.

Keywords: Location based services · Social networks · Electronic commerce

J. C.-W. Lin et al. (Eds.): MISNC 2019, CCIS 1131, pp. 155–163, 2019.
https://doi.org/10.1007/978-981-15-1758-7_13

1   Introduction

With the rise of social networks in recent years, a number of researchers have begun to discuss how social networks can be leveraged to improve response time and response measures during disasters. The idea behind this is that social networks allow the task of information collection to be crowdsourced, which allows officials and disaster responders to obtain a more complete picture of a disaster immediately after it occurs. Some examples of research into social networks and disaster mitigation are as follows. In 2010, Goodchild and Glennon [1] presented a method that employs crowdsourcing to ascertain the scope of a disaster. The following year, Gao *et al.* [2] proposed a method that employs crowdsourcing to coordinate disaster relief efforts. In 2012, Chu *et al.* [3] gauged the scope of disasters using Web 2.0. The year after that, researchers in Japan developed a set of algorithms to better understand flood disasters in the Kyoto area based on Twitter Tweets [4]. Similarly, Lai and Yang [5] proposed a social network-based public notification system for disaster management. In 2014, Hsueh *et al.* [6] used Facebook posts to understand post-rain flooding conditions in various areas. More recently, Wukich [7] employed social media to manage disaster-related issues. These studies emphasize the great help that social networks can provide in identifying disasters, obtaining comprehensive information about disasters, and even in managing disasters.

However, despite many successful crowdsourcing-based analyses and algorithms, to date, no disaster management system based on crowdsourcing has been created in Taiwan. We speculate that there are three reasons for this. (1) If the information is being crowdsourced from existing social networks, privacy settings may prevent data collection, and formatting issues may affect data processing. Furthermore, if the social network requires modifications, the entire information retrieval program that it employs must be redesigned from the ground up. Additionally, existing location-aware functions of social networks are only auxiliary tools; therefore, results of disaster-related analyses can be severely impacted by a lack of geographic information. (2) Many researchers and organizations have attempted to construct their own platforms to collect crowd-sourced information, but these platforms were only designed for disaster events, so users did not use them in their everyday lives. By the time a disaster occurred, users had long forgotten about the existence of these platforms and did not use them to report disaster information. Clearly, the leveraging of proprietary social networks is challenged by a lack of information. (3) Another important problem is the lack of information in some areas. From a theoretical perspective, using information from social networks to assist in disaster relief is very reasonable. If people located around a disaster area are able to provide relevant information, the time required to complete an analysis should be greatly reduced, and the results should be reasonable. Currently however, users provide information via social networks by their own accord; they cannot be forced to provide it. Resultingly, in some areas there may be many people uploading information about a disaster to social networks, while in other areas there may be none, as shown in Fig. 1. In that figure, a large residential area is located to the south of the disaster area, while a large, empty lot lies to the east. In the residential area, many users can provide information via social networks. However, no users are available to upload information from the empty lot. This impacts the ability of officials and first responders to obtain comprehensive information about the disaster. Clearly, something should be done to help government authorities to ask nearby users about disasters is the solution to this problem.

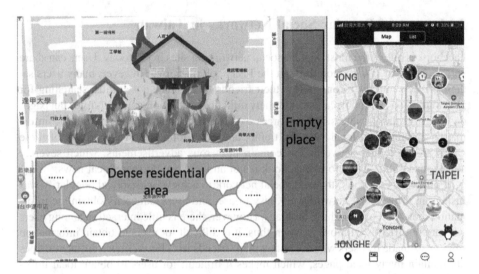

Fig. 1. An example of the lack of information in some area.   **Fig. 2.** Interface of the proposed social network

This study proposes a novel, location-aware social network to overcome the aforementioned shortcomings. This location-aware social network is an extension of previous work [8]. The operating concept of our proposed social network differs from those of conventional mainstream social networks, as it focuses on geographic space. Every post has a GPS location attached to it, and posts are viewed on a map, as shown in Fig. 2. Users can only see posts from areas near where they live or close to their current location. We believe that this design approach will allow our proposed social network to help users better understand issues affecting their surrounding environment. This should in turn encourage user base growth, eventually leading to a large number of users. In the event of an disaster, disaster management agencies will also be able to obtain and analyze text, sounds, images, and videos from locations around the disaster using the backend of our social network. Finally, our proposed social network also offers a channel through which disaster management agencies can make disaster posts in order to interact with residents living near the location of the disaster and thereby understand the situation from multiple perspectives.

In developing the proposed social network, we added a number of interactive functions that can be employed by management agencies and users to perform disaster-related analyses more effectively. We also included commercial promotion features and commercial analysis functions, so that we can increase the number of users. Moreover, we added an altered system framework to provide management agencies with a convenient platform that facilitates information retrieval.

The content of this paper includes the following. Section 2 presents a review of popular social networks in Taiwan; Sect. 3 introduces the functions of the proposed social network; Sect. 4 explains the system framework; and Sect. 5 contains the conclusions.

2 Review of Popular Social Networks in Taiwan

Facebook is one of the most popular social networking sites in use today. Users can share various types of information through the Facebook platform. However, many users set their privacy settings to only allow their friends to see their posts; thus, searching for information on Facebook can be extremely difficult. In addition, many clubs or fan pages dedicated to daily matters in local areas exist on Facebook. These pages share information pertaining to local weather, traffic, and other topics. However, many of these pages are no longer active and/or contain numerous advertisements, which severely affects readability. Therefore, Facebook is not an ideal way to share local news.

Following Facebook, the next social network to achieve widespread popularity was Instagram. This social network very easy to use, and users can easily obtain the information they want by searching hashtags. However, there is little interaction among users who are not friends with one another; thus, it is difficult for users to obtain information from posts made by users who are not on their 'friends list'. Furthermore, Instagram does not contain many local pages, which makes it difficult for users to obtain local information from the platform. In other words, this social network has almost no capacity to help users obtain information involving the geographic space they are located in.

Twitter is another popular social network. However, it has more users abroad and is less common in Taiwan because it has a more complex interface and more complicated functional procedures. Thus, obtaining local information pertaining to Taiwan on Twitter is fairly difficult. Indeed, Twitter itself does not provide local information; it focuses on posts centered on people. Resultingly, this social network platform is not suitable for disaster management either.

Dcard is the social network platform most commonly used by university students in Taiwan. It categorizes posts by university. Many university students post on this platform, and the interface is easy to use. Therefore, it is fairly easy to obtain information using Dcard. However, it is also easy to open an account on this social network, and posts can be made semi-anonymously. Resultingly, fake posts are common, which means that an analysis based on Dcard posts would yield findings of questionable accuracy.

PTT is a bulletin board system used by many individuals between the ages of 20 and 30 in Taiwan. A number of Taiwanese studies have been based on this social network. Registration is completely anonymous but involves a complex procedure. In addition, the operation and functions of this platform are fairly complicated. Nonetheless, PTT provides users with a comfortable reading environment. The information contained on PTT tends to be quite accurate; however, it is often not local or continuously updated. Therefore, whether it can be leveraged to improve disaster management efforts is still under debate.

Pixnet is a social network used by bloggers in Taiwan. It is mainly used for longer articles but contains many advertisements, which affects the ease of reading.

Nonetheless, this social network contains an abundance of highly reliable local information which bloggers frequently update. However, articles are categorized by blogger, and it is difficult to search for information pertaining to specific areas.

3   Functions of Interactive, Location-Aware Taiwanese Social Network

The interactive social network developed by our research team includes three location-aware functions: (a) a map presentation of all information on the social network, (b) post and comment screening, and (c) personal memory maps. Our team also developed a commercial version of the proposed social network which includes (d) a commercial promotion and analysis model. In addition, the commercial model includes four inter-active functions for disaster management: (e) a user disaster distress alert, (f) disaster notifications for disaster management agencies, (g) disaster notifications for users, and (h) nearby status queries from disaster management agencies. Functions (a), (b), and (c) are similar to those of the Animap platform that our research team previously developed [8], while functions (d) through (h) were added to increase the number of users who sign up for the social network, thereby making it more useful to disaster management efforts. The details of each function are described over subsequent sections:

(a) Map presentation of all information on the social network: An example of posts and comments as they appear on our proposed social network is shown in Fig. 2. The posts and comments are displayed on a map based on the user's geographic location, thereby showing the latest local news and nearby landmark information. This information can be shared by nearby users, enabling others to easily obtain local information.

(b) Post and comment screening: Our proposed social network displays posts and comments on maps. If there are too many people flowing through an area, there may be too many comments left by people there, resulting in users seeing too many posts and comments at the same time. For cases such as these, our social network features screening settings that allow users to quickly find the content they are interested in, as shown in Fig. 3.

(c) Personal memory map: The personal memory map function of our social network is shown in Fig. 4. This feature allows users to store their previous posts as well as locations they have visited before. This in-turn allows users to archive happy memories about where they have been in the past.

(d) Commercial promotion and analysis model: Revenue from commercial advertisements are crucial to the operations of a social network. Accordingly, we developed corresponding functions for commercial purposes. These functions allow businesses to upload promotional videos about their products. The videos appear in the social network of users who are located near the business in hopes of attracting them. However, to ensure that an excess of advertisements do not flood the social network and affect its usefulness, businesses will be limited to uploading one video per day. Ultimately, our commercial model will also provide businesses with analytics pertaining to user locations and viewing records. This should allow businesses to better understand who their potential customers are and how many users are viewing their advertisements. In other words, analysis results should allow businesses to better understand user needs and/or enhance advertisement quality in such a way as to attract more viewers.

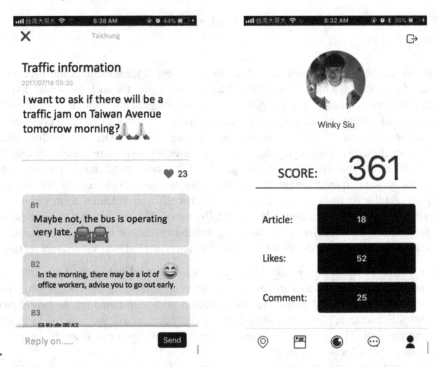

Fig. 3. Example of post and comment screening   **Fig. 4.** Personal memory map

(e) User disaster distress alert: Our proposed social network emphasizes localization and supports the long-term care concept of community residents helping one another. Accordingly, our social network features an disaster distress icon. When a user taps the icon, nearby users receive an disaster distress alert which notes the location from which the alert was sent. This information should permit any willing user to go to the disaster location and provide assistance, as shown in Fig. 5.

(f) Disaster notifications for disaster management agencies: When an disaster occurs, typically, many users who are near the disaster location generate posts at the same time. When the system detects numerous posts which (1) contain similar previously flagged words and (2) originated from a single area, it sends an disaster notification to disaster management agencies so that these agencies can determine whether an disaster has occurred and respond accordingly. (Note that, for the sake of professionalism, all words which are flagged in our social network were provided and defined by disaster management agencies prior to the release of the network.) For example, if numerous posts containing flagged words such as "smoke", "fire", or "burning smell" appear in an area, the system will notify disaster management agencies. Conversely, if numerous posts containing unflagged words such as "movie filming", "so hot", or "come quick" appear in an area, disaster management agencies will not be notified. (The post frequency threshold for disaster management agency notification was also provided by the disaster management agencies themselves.)

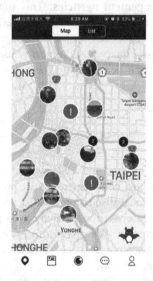

Fig. 5. Example of user disaster distress alert

Fig. 6. Example of disaster notifications for users

Fig. 7. Nearby status query from a disaster management agency

(g) Disaster notifications for users: The purpose of this function is to keep nearby users a safe distance away from disaster events such as fires or gas explosions. Once the approximate range of an disaster event has been determined, disaster management agencies can use this function to send notifications to nearby users so that they can avoid the site of the disaster and take appropriate response measures, as shown in Fig. 6.

(h) Nearby status queries from disaster management agencies: This function enables disaster management agencies to interact with users who are located near the disaster event. Specifically, after an disaster occurs, disaster management agencies can use our social network to send disaster query posts to nearby users so that they can gain more comprehensive information pertaining to the disaster, as shown in Fig. 7.

4   System Structures

The structure of our social network is divided into a frontend and a backend, as shown in Fig. 8(a). The frontend includes the map of posts and comments, popular post and comment recommendations, the comment function, local discussion boards, and personal home pages. The backend involves the use of Firebase for the analysis of posts, comments, multimedia information, and user information. We opted to use Firebase because this database supports text and multimedia storage space simultaneously, which greatly sped up the development of our social network.

In addition, users are able link to the server directly. This does not create issues when the number of online users is small; however, when many users are logged on to the network, there is a risk that the system could crash. We therefore applied the method displayed in Fig. 8(b) to connect users to the server. In brief, users first link to the platform server, and the platform server then connects them to both the backend database and the overall server. This hierarchical diversion should prevent the server from crashing when there are too many users logged on to the social network.

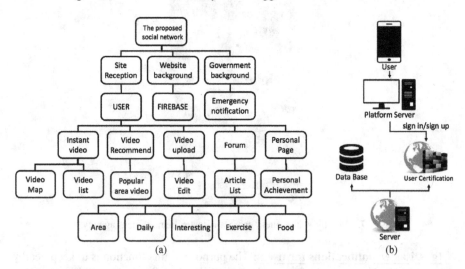

Fig. 8. System structure employed by our social network: (a) frontend and backend structure; (b) flowchart illustrating how users connect to the server.

5 Conclusions

With the flourishing development of social networks, an increasing amount of attention is being given to how social network crowdsourcing can be leveraged to benefit disaster management. However, the realization of these benefits is currently challenged by limitations in the functionality of existing social networks, and the public has shown a lack of interest in becoming users of self-developed platforms designed to better facilitate crowdsourcing. Consequently, no social network applications have been successful in leveraging crowdsourcing to benefit disaster management thus far. To address these limitations, we developed an interactive, location-aware social network. We believe that our proposed network should attract a large and dedicated base of daily users because it includes features and functionality which differ from those of existing social networks. Upon achieving such a user base, it will be possible for our network to facilitate the crowdsourcing of information following an disaster, thereby improving the disaster response capabilities of government authorities.

Acknowledgments.   This work was supported in part by the Ministry of Science and Technology of Taiwan, R.O.C., under Contracts MOST 107-2625-M-224 -003 and MOST 107-2119-M-224 -003 -MY3.

References

1. Goodchilda, M.F., Glennona, J.A.: Crowdsourcing geographic information for disaster response: a research frontier. Int. J. Digit. Earth **3**(3), 231–241 (2010)
2. Gao, H., Wang, X., Barbier, G., Liu, H.: Promoting coordination for disaster relief – from crowdsourcing to coordination. In: Salerno, J., Yang, S.J., Nau, D., Chai, S.-K. (eds.) SBP 2011. LNCS, vol. 6589, pp. 197–204. Springer, Heidelberg (2011). https://doi.org/10.1007/978-3-642-19656-0_29
3. Chu, E.T.-H., Chen, Y.-L., Lin, J.-Y., Liu, J.W.: Crowdsourcing support system for disaster surveillance and response. In: Proceeding on International Symposium on Wireless Personal Multimedia Communications (2012)
4. Sato, H., Takeda, K., Matsumoto, K., Anai, H., Yamakage, Y.: Efforts for disaster prevention/mitigation to protect society from major natural disasters. Fujitsu Sci. Tech. J. **52**(1), 107–113 (2016)
5. Lai, D.C., Yang, I.T.: Public disaster mitigation system based on social network and information push technology, Master thesis of National Taiwan University of Science and Technology (2013)
6. Kat, M.W., Hsueh, N.L., Chen, Y.C.: An opinion mining approach based on localized social group and query expansion, Master thesis of Feng Chia University (2016)
7. Wukich, C.: Social media use in emergency management. J. Emerg. Manag. **13**(4), 281–294 (2015)
8. Yao, S.: ANIMAP- a location-based social networking services. In: Proceeding on IEEE International Conference on Applied System Innovation (2019)

A Fast Approach of Graph Embedding Using Broad Learning System

Long Jiang[1], Yi Zuo[1(⊠)], Tieshan Li[1], and C. L. Philip Chen[1,2]

[1] Navigation College, Dalian Maritime University, Dalian, Liaoning, China
{jianglong, zuo}@dlmu.edu.cn, tieshanli@126.com,
Philip.Chen@ieee.org
[2] Faculty of Science and Technology, University of Macau, Macau, China

Abstract. In this paper, traditional DeepWalk method and broad learning system (BLS) are used to classify network nodes in graph embedding, and results of classification are compared. When categorizing, DeepWalk adopts one vs rest (OvR) logistic regression method, and BLS is applied after the production of vector representations. In order to obviously compare results of the two classification methods, Support Vector Machine (SVM) and Convolutional Neural Network (CNN) are employed to carry out the experiment on multi-label classification of BlogCatalog. The experimental result shows that F1 score of BLS is obviously higher than DeepWalk and other methods, and training time of BLS is much less than other methods. These performances make our method suitable to graph embedding.

Keywords: DeepWalk · Graph embedding · Network representation · Multi-label classification · Broad learning system

1 Introduction

A graph is a highly abstract and expressive data structure that describes the association between entities and entities by defining nodes and edges. Commonly used diagrams include social networks, commodity networks, knowledge maps, and etc. Edges and nodes of the network graph usually contain rich information. Graph embedding is an approach to learning latent representations of nodes in the form of vectors, and it encodes the association relationship of nodes in a continuous vector space, which is convenient for calculating the association relationship between nodes.

DeepWalk is a method to learn latent representations of a graph's vertices [1]. It applies the principle of SkipGram, using a vertex to predict vertices before and after it by random walk, and learns to obtain the latent representation of the vertex. It adopts the relational classification to obtain the structure of the graph first, and then using the structure of the graph to perform unsupervised, label-independent method to achieve classification, the classifier of DeepWalk is a linear classifier (logistic regression). The core of traditional DeepWalk is DeepWalk (DW) procedure and one vs rest logistic regression, which are mentioned in Sect. 2. The overview of DeepWalk is shown in Fig. 1.

© Springer Nature Singapore Pte Ltd. 2019
J. C.-W. Lin et al. (Eds.): MISNC 2019, CCIS 1131, pp. 164–172, 2019.
https://doi.org/10.1007/978-981-15-1758-7_14

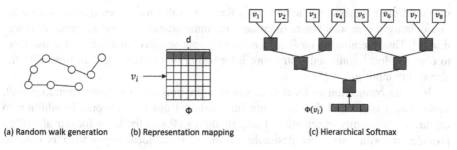

Fig. 1. Overview of DeepWalk

Broad learning system (BLS) is a newly developed network framework that provides an alternative to learning deep structures without the need for time-consuming training procedures [2]. BLS has a high efficiency on avoidance of numerous parameters of the multi-layer network and the fine-tuning steps based on backpropagation. Therefore, it performs better than many algorithms in multi-label classification. There are two main contributions in this paper. First, we introduce broad learning system to solve multi-label classification problem of the graph using a fast approach. Second, we evaluate our method on the social network of BlogCatalog, and our classification performance gets improvements over 10% of Micro-F1, and the training time of BLS is less than 10% of other methods.

The rest of the paper is arranged as follows: In Sect. 2, we give the literature review of graph embedding and an introduction of multi-label classification and its related methods. In Sect. 3, we sketch our experiments and discuss the experimental results. We close with our conclusion and future work in Sect. 4.

2   Literature Review

2.1   Review of Graph Embedding

We consider the issue of classifying members of a social network into one or more categories. Let $G = (V, E)$, where V is the network member, E is their connection, $E \subseteq (V \times V)$, $G_L = (V, E, X, Y)$ is a partially marked social network, attribute $X \in R^{|V| \times S}$ where S is the size of the feature space of each attribute vector, and $Y \in R^{|V| \times |y|}$, y is the label set.

Social network refers to relatively stable relationship system formed by the interaction between individual members of society. The social network focuses on the interaction and connection between people, and social interaction will affect people's social behavior. A social network is a social structure composed of many nodes. A node usually refers to an individual or an organization. The social network represents various social relationships, including friend relationships, classmate relationships, business partnerships, ethnic belief relationships, and etc.

We denote the random walk from a root vertex v_i as w_{v_i}. It is a stochastic process of producing the random variables, $w_{v_i}^1, w_{v_i}^2, \cdots, w_{v_i}^k$, where $w_{v_i}^{k+1}$ is a vertex randomly selected from neighbors of the vertex v_k. Random walks have been used as a measure of similarity for various issues in content recommendation [3] and community detection [4]. They are also the basis for a class of output-sensitive algorithms that use them to compute local community structure information that is consistent in time with the size of the input graph [5].

It is the connection to local structure that prompts us to use short random walk streams as a basic tool for extracting information from the network. In addition to capturing community information, using random walks as the basis for our algorithm provides us with two other desirable attributes. First, local exploration is easy to parallelize. Several random walkers (in different threads, processes, or machines) can explore different parts of the same graph at the same time. Second, relying on information obtained from short random walks makes it possible to adapt to small changes in the graph structure without global recalculation. We can iteratively update the learning model, where new random walk is linear in time from the changing region to the entire graph.

2.2 Multi-label Classification and Related Methods

Given a social network and knowing the behavior label information of some users (nodes), our task is to classify the behavior labels of other unclassified nodes in the network. Suppose all the behaviors in a network can be classified into K labels $\{c_1, c_2, \cdots, c_k\}$, each with a value c_i of 1 or 0. For example, a user may join multiple interest groups, $c_i = 1$ indicates that the user has joined interest group, otherwise $c_i = 0$. Our research questions can be formally stated as follows:

Suppose there are K labels $y = \{c_1, c_2, \cdots, c_k\}$, given a network $G = \{V, E, Y\}$, where V is the set of vertices, E is the set of edges, $Y_i \subseteq y$ is class label of the vertex $v_i \in V$, and Y_i values of some vertices $v_i \in V^L (V^L \subseteq V)$ are known, how do we infer values of remaining vertices (or probability distribution for each label) is the problem of multi-label classification.

There are many methods of multi-label classification, such as logistic regression, SVM and CNN. DeepWalk adopts the one vs rest logistic regression, and the thought of one vs rest is: When n types of samples are classified, one sample is taken as one class, and the remaining samples of all types are treated as another class, thus forming n binary classification problems, using logistic regression algorithm for n times, the data set trains n models, and samples to be classified are inputted into n models, and the label of sample corresponding to the model with highest probability is considered to be label of the sample.

2.3 Broad Learning System

Broad learning system is designed in accordance with the idea of Random Vector Functional-link Neural Network (RVFLNN), and Figs. 2 and 3 show the standard structures of RVFLNN and BLS in the form of a functional link neural network [6].

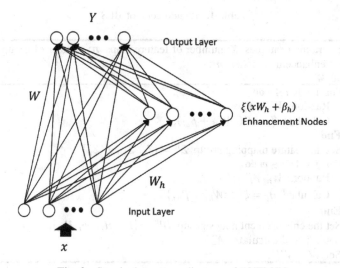

Fig. 2. Standard structure diagram of RVFLNN

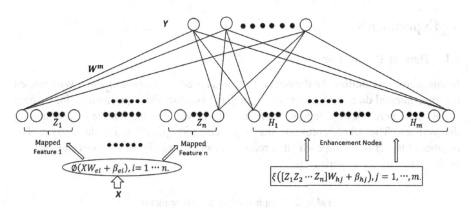

Fig. 3. Standard structure diagram of BLS

Strictly speaking, the input is converted into a random feature represented as a "feature node" and the structure is further extended in the manner of "enhancement nodes". In other words, the input data is first converted to a random feature by a given feature mapping function. The resulting feature is then further connected to the enhancement node by a nonlinear or linear activation function. Finally, the random node is fed directly to the output along with the enhancement node. The weight of the output layer is determined by an effective pseudo-inverse approximation (ridge regression) or backpropagation based on a gradient of the least square cost function. Table 1 displays the pseudo code of BLS.

Table 1. Pseudo code of BLS

Input:	training samples X; number of feature maps n; number of groups of enhancement nodes m;
Output: W	

1. **For** $i = 0; i \leq n$ **do**
2.   Random W_{ei}, β_{ei};
3.   Calculate $Z_i = \emptyset(XW_{ei} + \beta_{ei})$;
4. **End**
5. Set the feature mapping group $Z^n = [Z_1, Z_2, \cdots, Z_n]$;
6. **For** $j = 1; j \leq m$ **do**
7.   Random W_{hj}, β_{hj};
8.   Calculate $H_j = \xi(Z^n W_{hj} + \beta_{hj})$;
9. **End**
10. Set the enhancement nodes group $H^m = [H_1, H_2, \cdots, H_m]$;
11. Set A_n^m and calculate $(A_n^m)^+$
12. Set $W = W_n^m$;

3 Experiments

3.1 Dataset Description

In this section we outline the dataset and methods used in our experiments. We selected the experimental dataset of BlogCatalog in [7] to conduct the experiment. BlogCatalog is the social network of bloggers. There are 10,312 vertices and 333,983 edges among the vertices. The label represents the topic category provided by the author. The number of labels is 39 and each label represents an interest. Table 2 shows an overview of the graph in our experiments.

Table 2. Graph used in the experiments

Name	BlogCatalog
$|V|$	10,312
$|E|$	333,983
$|y|$	39
Labels	Interests

3.2 Experimental Analysis

3.2.1 Macro-F1 and Micro-F1

In order to calculate Macro-F1 and Micro-F1, first, we need to introduce four basic concepts: True Positive (TP), False Positive (FP), False Negative (FN) and True Negative (TN).

- TP: The forecast is positive, and the actual is positive;
- FP: The forecast is positive, and the actual is negative;
- FN: The forecast is negative, and the actual is positive;
- TN: The forecast is negative, and the actual is negative.

Then, we can calculate the precision and recall by the following formula:

$$Precision = \frac{TP}{TP + FP},$$ (1)

$$Recall = \frac{TP}{TP + FN},$$ (2)

Next, we can calculate F1 score, F1 score is an indicator used in statistics to measure the accuracy of a binary model. It takes into account precision and recall of the classification model. The F1 score can be thought as a weighted average of the accuracy and recall, with a maximum of 1 and a minimum of 0. The formula of F1 score is as follows:

$$F1 = 2 \cdot \frac{Precision \cdot Recall}{Precision + Recall},$$ (3)

Finally, in a multi-label classification problem, in addition to calculating F1 for each class, we also need to combine F1 of all classes for consideration, and there are two methods to merge them:

The first one is called Macro-F1, and the formula of Macro-F1 is:

$$MacroF1 = \frac{2 \sum_{i=1}^{m} TP_i}{2 \sum_{i=1}^{m} TP_i + \sum_{i=1}^{m} FP_i + \sum_{i=1}^{m} FN_i},$$ (4)

The second one is called Micro-F1, and the formula of Macro-F1 is:

$$MicroF1 = \frac{1}{m} \sum_{i=1}^{m} \frac{2TP_i}{2TP_i + FP_i + FN_i},$$ (5)

3.2.2 Experimental results

In order to see the superiority of BLS, in addition to OvR, we also compare this algorithm with two baseline methods: CNN and SVM. We utilize the dataset and experiment process in [7] and report the performance in with the form of Macro-F1 and Micro-F1.

First, we employ DW to produce representation vectors. Then, input these vectors to different models, including BLS, CNN and SVM. Finally, the outputs and labels are compared and macro-F1 and micro-F1 are calculated.

The Macro-F1 and Micro-F1 of our experiments are presented in Table 3 and Fig. 4. From the table, we can see that BLS performs better than OvR, CNN and SVM. In fact, the Macro-F1 of BLS is over three times as OvR and SVM. In addition, the Micro-F1 of BLS is also more than two times of traditional DeepWalk in value.

Table 3. Graph used in the experiments

	Macro-F1(%)	Micro-F1(%)
DeepWalk(DW+OvR)	29.41	45.24
BLS(DW+BLS)	92.68	100
SVM(DW+SVM)	30.56	100
CNN(DW+CNN)	81.27	83.36

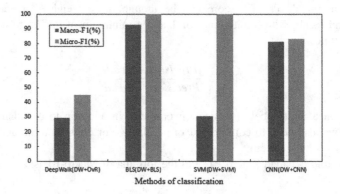

Fig. 4. Results of multi-label classification in BlogCatalog

Table 4. Results of training time in BlogCatalog

	Training time(s)
DeepWalk	100.58
BLS	9.23
SVM	229.06
CNN	565.54

The training time of our experiments are presented in Table 4 and Fig. 5. From the experimental results, we can see that BLS also performs better than DeepWalk, CNN and SVM. In fact, the training time of BLS is less than 10% of DeepWalk. In addition, Although SVM performs the same as BLS in Micro-F1, and the Macro-F1 and Micro-F1 score of CNN and BLS are close, their training time is much longer than BLS.

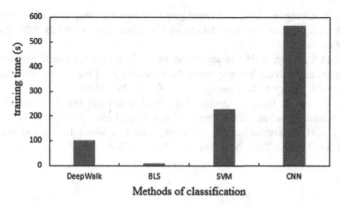

Fig. 5. Results of training time in BlogCatalog

4 Conclusion

We introduce an effective approach of Broad Learning System, for multi-label classification of graph embedding. Using the technique of DW, our method classifies vertices in social network more accurate. Experiments of BlogCatalog indicate the effectiveness of our method in multi-label classification for large-scale graph. Our future work will concentrate on the optimal method of multi-classification on large-scale graph, such as Flickr or Youtube, since the performance of BLS in BlogCatalog outperforms other methods obviously, we can expect the classification of vertices in larger graph will be better.

Acknowledgments. This work is supported in part by the National Natural Science Foundation of China (under Grant Nos. 6175202, 61751205, 61572540, U1813203, 61803064, 71831002); the LiaoNing Revitalization Talents Program (under Grant No. XLYC1807046); the Science & Technology Innovation Funds of Dalian (under Grant no. 2018J11CY022); the Program for Innovative Research Team in University (IRT_17R13) and the Fundamental Research Funds for the Central Universities (3132019501, 3132019502).

References

1. Perozzi, B., Al-Rfou, R., Skiena, S.: DeepWalk: online learning of social representations. In: Proceedings of the 20th ACM SIGKDD International Conference on Knowledge Discovery and Data Mining, pp. 701–710 (2014)
2. Chen, C.L.P., Liu, Z.L.: Broad learning system: an effective and efficient incremental learning system without the need for deep architecture. IEEE Trans. Neural Netw. Learn. Syst. **29**(1), 10–24 (2017)
3. Fouss, F., Pirotte, A., Renders, J.-M., Saerens, M.: Random-walk computation of similarities between nodes of a graph with application to collaborative recommendation. IEEE Trans. Knowl. Data Eng. **19**(3), 355–369 (2007)

4. Andersen, R., Chung, F., Lang, K.: Local graph partitioning using pagerank vectors. In: 47th Annual IEEE Symposium on Foundations of Computer Science, FOCS 2006, pp. 475–486. IEEE (2006)
5. Spielman, D.A., Teng, S.-H.: Nearly-linear time algorithms for graph partitioning, graph sparsification, and solving linear systems. In: Proceedings of the Thirty-Sixth Annual ACM Symposium on Theory of Computing, pp. 81–90. ACM (2004)
6. Chen, C.L.P., Liu, Z., Feng, S.: Universal approximation capability of broad learning system and its structural variations. IEEE Trans. Neural Netw. Learn. Syst. **30**, 1–14 (2018)
7. Tang, L., Liu, H.: Relational learning via latent social dimensions. In: Proceedings of the 15th ACM SIGKDD, KDD 2009, New York, USA, pp. 817–826 (2009)

A Study on Pricing and Diffusion of New Services – an Experimental Research on Electronic Books Market

Masashi Ueda[✉]

Doshisya University, Shokokuji-Monzen-Cho, Kamigyo-Ku,
Kyoto 602-8580, Japan
skl15573@mail.doshisha.ac.jp

Abstract. We treat distribution matters of Japanese electronic book service in this article. In this context the relationship between distribution factors and models are so important. In this paper we compare this issue with electronic money cards. Then review our research analysis from the view point of this prospect.

Keywords: Service adoption · Distribution theory · e-distribution

1 Introduction

In some countries diffusion speed of new services is relatively slow than other countries though physical conditions for infrastructure is enough. As Rogers (1962) innovation has stages; innovators (2.5%), early adapters (13.5%), early majorities (34%), late majorities (34%), and laggards (16%). In our previous works we found a tendency of diffusion patterns of e-distribution markets in Japan by collection of consumer data via web survey and analysis them. In this paper we focused on this contradiction. We analysed the consumer behaviour in the e-book market in Japan. In 2009 Japanese e-book market was third times as that of the U.S. but afterward the U.S. market expanded rapidly, but the growth of the Japanese market was limited. Partially because Amazon's Kindle shock in the U.S. but we think other factors may exist.

2 Electronic Money Cards and Electronic Books

And according to BOJ (2019) electronic money is widely accepted in Japanese society. And both volume of transactions and value of transactions are increasing constantly. This is one of the main payment methods for cashless payment. Now one person has 2.9 electronic money cards in Japan in 2017. But most of them is stowed in smart card based on a prepaid payment instruments called as FeliCa.

When we look at the diffusion of electronic books, we will overview the diffusion process of electronic money that had done in advance as showed in BOJ data.

© Springer Nature Singapore Pte Ltd. 2019
J. C.-W. Lin et al. (Eds.): MISNC 2019, CCIS 1131, pp. 173–179, 2019.
https://doi.org/10.1007/978-981-15-1758-7_15

According to Okada (2018) there are the following factors in the diffusion of electronic money cards in Japan.

Table 1. Latest e-money market in Japan

	Vol. of Tran. (mil.)	Val. of Tran. (100 mil. Yen)	# of Terminals (10,000)	Val. Outstanding (100 mil. Yen)
2016	5,192	51,436	199	2,541
2017	5,423	51,994	230	2,747
2018	5,853	54,790	273	2,975

Source: BOJ (2019)

And he reported stakeholders also thought that these four factors are key concepts for adoption of electronic money cards in Japan. In late 1990s, it was completely for technology geeks, as it had limited space available. He survived one week only with 'supercash', an experimental e-money in 1999 in Shinjuku (Table 2).

Table 2. Key concepts for e-money cards.

Novelty (新奇性)	The pleasure of touching innovation.
Convenience (利便性)	Convenience to touch and pass through the ticket gate.
Profitability (利得性)	Earn points, increase discount rate.
Publicity (公共性)	Leads to the revitalization of local communities.

Source: Okada (2018)

And he reported stakeholders also thought that these four factors are key concepts for adoption of electronic money cards in Japan. In late 1990s, it was completely for technology geeks, as it had limited space available. He survived one week only with 'supercash', an experimental e-money in Shinjuku in 1999.

Afterward smart traffic card Suica, implemented the e-money function. in 2004, was introduced in JR East in November 2001. It was more convenient than traffic magnetic cards or paper tickets. So, it was the first smart cards widely accepted in

In same month Edy!, an early e-money operated by SONY subsidiary company, also released. It held many point-up programs with shopping streets around the Japan. These can be said profitable.

One of the latest one can be WAON, operated by biggest supermarket chain AEON, in 2007. Though AEON is national wide chain store but had need to co required local cooperation in various places because it was thought to be outsider for each local community. So, Aeon introduced a mechanism to donate to local projects if you use your local WAON. These can be said publicity (Table 3).

Table 3. Diffusion stages and supposed key concepts.

Innovators (2.5%)	Fun (娯楽性)
early adopters (13.5%)	Novelty (新規性)
early majorities (34%)	Convenience (利便性)
late majorities (34%)	Gain (利得性)
laggards (16%)	Ppublic interest (公共性)

Source: Ueda (2018)

These lead to very similar results. These three studies can be summarized in the following table.

Table 4. Diffusion stages and supposed key concepts.

Rogers	Ueda	Okada
Innovators (2.5%)	Fun (娯楽性)	Novelty (新奇性)
early adopters (13.5%)	Novelty (新規性)	
early majorities (34%)	Convenience (利便性)	Convenience (利便性)
late majorities (34%)	Gain (利得性)	Profitability (利得性)
laggards (16%)	Public interest (公共性)	Publicity (公共性)

As previous survey on e-book market in Japan, Ueda (2017) note that 'though limitation of titles Japanese e-book market come to be mature in near future if the market structure will not change as it is'. This conclusion is a little bit different from CPRC (2013) that say Japanese e-book market is in early stage. So, in this paper we collect consumer data via web survey and analyse them and it reviles the tendency of Japanese consumer for e-distribution services.

3   Review of Collected Data in 2016

Most of e-book are sold via the Internet in Japan. So, sampling bias may be limited if we utilize online survey. Our data was collected by the Internet questionnaires survey. Here is the outlook of Web survey shown by Table 1. This data was collected in 2016 (Table 5).

Table 5. Outlook of web survey

Surveillance period		From 25th to 27th of March 2016
Survey's company		Rakuten Research
Total samples		3,000
Gender		Female: 1,500 and Male: 1,500
Age groups	To 20 s	(F: 150, M:150),
	30 s	(F: 150, M:150),
	40 s	(F: 150, M:150),
	50 s	(F: 150, M:150),
	over 60 s	(F: 150, M:150).

Source: Ueda (2017)

Our samples are monitors selected by age group and sex in equal spacing from Rakuten Research. Only about 10% of them have an experience using certain e-book services though 36% of them use applications over their smartphones.

After we check the WTP for e-book simply we need to check real consumers' preference by using choice type conjoint analysis.

We set six cards for selection in five factors for illustrate e-book services; price, number of titles, point back services, devices, providers.

In details are shown in Table 6 and a sample set is also shown Table 7. We provided six cards for choice and respondents select one from them and each respondent answered 12 sets of selections.

Table 6. Factors and their standards for conjoint

Price	1,000 Yen	800 Yen	500 Yen
# of Titles	1 million	0.5 million	
Point back	5%	1%	Non
Devices	Smartphones (S) and Tablets (T)	S, T, Web browsers (W), and e-book readers (R)	–
Providers	Small Ventures (V)	Major Established (M)	–

Table 7. A sample set of choice cards

	I	II	III	IV	V	VI
Price (10Yen)	100	100	100	50	80	80
Titles (10,000)	100	50	100	100	100	100
Point back (%)	0	1	5	1	1	1
Devices	S	S	S	S	S	S
	T	T	T	T	T	T
		W	W	W		W
		R	R	R		R
Providers	V	M	V	M	M	M

And we check the WTP for e-book simply we need to check real consumers' preference by using choice type conjoint analysis (Table 8).

We set six cards for selection in five factors for illustrate e-book services; price, number of titles, point back services, devices, providers. We calculate coefficient of each parameters and marginal willingness to pay for four factors. According to Table 4 each factor is in significance level by t values and p values. As Ueda (2017) said number of tiles has very week tile with sales, our result also very limited MWTP for tiles. As 1. Atchariyachanvanich et al. (2008) pointed out that Japanese consumer tend to collect points issued by companies, but in our survey, point also limited MWTP.

Table 8. Conjoint outlook

	Coefficient	t values	p values	MWTP
Price (Yen)	-0.0313	-57.694	0.00	-
Titles (10,000)	0.0090	20.675	0.00	0.2886
Point Back (%)	0.0036	12.634	0.00	0.1155
Devices	0.3591	17.661	0.00	11.4740
Providers	0.4585	27.106	0.00	14.6500

Sample Numbers	36,000
Logarithmic likelihood	-56436.7

Source: Ueda (2016)

According to this data assuming that the MWTP of all consumers is linear, that preference indicates reliability to the service provider and high MWTP to e-book browsing on a multiscreen basis. This shows a high MWTP for convenience for the average consumer.

We estimated the discount rate consumers would like when changing from conventional to new products.

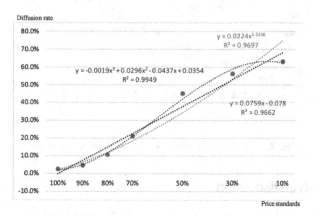

Fig. 1. Accumulated adopters in discount rate for e-books. Source: Ueda (2018a)

Figure 1 shows the relationship between price standards and discount rate. Theoretically price is diminishing, demand is increasing though marginal utility also diminishing. So, its shape would be serpentine curve or s-shaped curve. Impress Research Institute (2018) reported that Japanese e-book price standards are almost 80% of that of paper book. This means that about 30% gap is existing here and that this inhibits diffusion for late majorities (Table 9).

Table 9. Estimated relations

Linear	$y = 0.759x - 0.078$	$R^2 = 0.9662$
Power	$y = 0.0224x^{1.5256}$	$R^2 = 0.9697$
Polynomial	$y = -0.0019x^3 + 0.0296x^2 - 0.0437x + 0.0354$	$R^2 = 0.9949$

Source: Ueda (2018b)

Of these estimates, the one with the best fit is polynomial. This shape is very similar to the diffusion curve of the S-shaped curve presented by Rogers in the diffusion theory. Just innovator, early adapter, and early majorities are in 50% of the beginning. In addition, there is an inflection point at the boundary of 50%, which means that late majorities indicate that price elasticity is high.

Figure 2 shows the discount rate in the electricity market estimated similarly. In other words, we estimated how much the discount would shift to a new power company. Of these estimates, the one with the best fit is polynomial too. Similar diffusion curves were estimated for both markets.

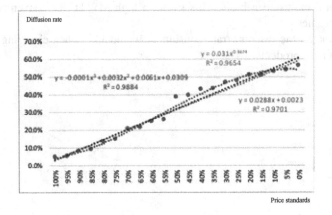

Fig. 2. Accumulated adopters of new Venture Elc. Comp.

4   Conclusive Discussion

In this paper, we examined the relation between Rogers's diffusion theory and the key concept in electronic money and the that in electronic book. As a result, the three showed very similar characteristics. In particular, it would be of some significance to indicate convenience and gain as data. On the other hand, we cannot present data on innovators and laggards, so we need to make certain reservations.

References

Atchariyachanvanich, K., Okada, H., Sonehara, N.: Theoretical model of internet shopping: evidence from a survey in Japan. Int. J. Electron. Customer Relat. Manage. 2(1), 16–33 (2008)

Bank of Japan: Payment and Settlement Statistics (May 2019), BOJ (2019)

Competition Policy Research Centre, Fair Trade Commission, Japan, A research on the Japanese e-book, CPRC CR 01-13 (2013)

Impress Research Institute: E-book business Research Report 2018, IRI (2018)

Okada, H.: People who made electronic money (3) Cashless Island, COMOMO (2018)

Rogers, E.M.: Diffusion of Innovations. Free Press of Glencoe, New York (1962)

Ueda, M.: A conjoint analysis on e-book market in Japan. In: ITS 2017 (2017)

Ueda, M.: A study on adoption of alternative services. In: Proceedings of MISNC 2018 (2018a)

Ueda, M.: Diffusion of e-Book in Japan from the view point of consumer activities. In: Proceedings of IDW18 (2018b)

Author Index

Printed in the United States
By Bookmasters